The Interdisciplinary Imperative

Interactive Research and Education, Still an Elusive Goal in Academia

List of Contributors

Arden Bement

Richard J. Brook

George Bugliarello

Robert Cahn

K.K. Deb

T. Egami

Rodney A. Erickson

Irwin Feller

Steward S. Flaschen

Tracy Gaudet

Scott Hauger

Lance Haworth

Michael Heylin

Joyce Jentoft

Mohammad Karim

Elton Kaufmann

Shigeyuki Kimura

Amitabha Kumar

Arunmugam Manthiram

Katy Marre

Paul C. Maxwell

Robert T. McGrath

Larry Murr

P.S. Nicholson

Forrest J. Remick

Rustum Roy

Manfred Rühle

Helmut K. Schmidt

Lyle Schwartz

Thomas Stoebe

Yuri Tretyakov

Andrew T. Weil

Robert Yager

Hiroaki Yanagida

The Interdisciplinary Imperative

Interactive Research and Education, Still an Elusive Goal in Academia

Rustum Roy, Editor

Writers Club Press
San Jose · New York · Lincoln · Shanghai

Published by Writers Club Press
an imprint of iUniverse.com, Inc.

For information address:
iUniverse.com, Inc.
620 North 48th Street, Suite 201
Lincoln, NE   68504-3467
www.iuniverse.com

ISBN: 0-595-01179-9

Printed in the United States of America

Contents ▸

· U.S. Programs

· Local

Foreword ▸

Executive Summary

The hundred participants in the two-day conference included distinguished individuals from all over the world representing all communities (universities, government, industry) each with long-term experience in the practice of "interactive" especially "interdisciplinary" research, specifically:

- *Leading scientists/engineers from academic, government and industrial research labs, worldwide, in the field of materials.*
- *Administrators from Universities and industry, typically Vice Presidents for Research or Department heads, and major funding agencies in the U.S. and in the major research countries: U.K., Germany, Japan, India, etc.*
- *Representatives of other scholarly fields (including the social sciences and medicine) where major interdisciplinary or interactive research/education thrusts have played or are playing a significant role.*

Definitions

The first genuine contribution of this conference was thoughtful attention to terminology. What is meant by a *"discipline?"* How is it related to the sum total of human knowledge? To the University's mission? Definitive papers provide excellent guidance here. The commonly used term *"interdisciplinary"* is equally indefinite. It has come to include a variety of activities in research and teaching which involve more than one unit from one institution. The loose use of such terminology is one of the causes for confusion in reports on the field. A more accurately inclusive term is "Interactive Research" which includes three different kinds of interactions:

1. Interaction across disciplines within one institution; Interaction across institutions (University-industry; university-government-laboratories; industry-government laboratories).
2. Interaction across research sectors from basic research through engineering to manufacturing within one, or across institutions.
3. Any combination of the above.

The relations among these kinds of interactions and the combinations thereof are most appropriately represented, as in the conference logo shown above, in one of the most familiar key concepts of materials research, the triaxial diagram.

Key Findings

1. There was unanimous agreement that virtually all cutting edge research in materials is, in fact, highly interactive (as defined above). The momentum towards greater and greater interaction in all three directions of I³R is enormous, and is dominated by industry's dealing with a completely new time-to-market parameter.

2. It has been shown empirically by the forty years of experience that *interactions across institutions*, university-industry, university-government, and even university-university are much easier than interactions within universities. Driven by industry's outsourcing of research and its proactive partnering with the best-matched Universities this IR branch is thriving. This was true in spite of certain barriers. (see Sec. 4 below)

3. In contrast, interdisciplinarity on a single campus is still struggling; there was general consensus that the University world had "not yet gotten its act together." After forty years *not one* good model was reported for a highly effective administrative structure within one institution, which actively advanced or supported (as distinct from permitted or tolerated) interdisciplinary research in *any* field (This is especially evident when there is no specific government contract requiring it). Indeed a former Commissioner of the U.S. Nuclear Regulatory Commission reported that on his former campus where he had been Assistant Vice President in charge of Interdisciplinary Research, 20 years ago, matters had gotten measurably worse.

   A senior manager of an eminent industrial research laboratory remarked that he was amazed that after so long "The University world simply hadn't 'gotten it': there is no other game in town other than interactive research."

4. Near unanimity was also evident in the deep dissatisfaction felt by both industrial representatives and active University researchers themselves with the present *intellectual property* arrangements offered by most Universities for university-industry joint work. (This same topic also pre-occupies the relevant committee of the Industrial Research Institute as the major issue in University-Industry relations.) There was agreement that the reason for this is the highly distorted perception by most University Presidents (most with virtually no background in the field), that Intellectual Property can be a major source of revenue for most Universities. The facts are, that less than a handful of Universities have obtained substantial (>$10 m) revenues from Intellectual Property. These are nearly all in the biomedical or software areas. The vast majority of University Intellectual Property offices cost more than their Intellectual Property income. However several industrial Vice Presidents and managers (and many faculty) drove home the point that the Universities *lost* many more millions of

immediate, certain, *research* dollars every year by turning away research partners from industry in the hope of a potential, maybe, uncertain, royalty streams 10-30 years downstream. As one senior manager of a major industrial giant put, it he finds it "much easier to do business with his competitors than with the Universities."

On the positive side some of the leading Universities in such inter-active research (e.g. MIT, Penn State, Arizona State) reported that *very* recently new models of intellectual property management had emerged, which held out some hope for real improvement (see Recommendations).

5. An important new finding was that a key factor almost always ignored in discussions of interdisciplinary research funding was the ***negative effects of the so-called "peer-review"*** system. Very senior U.K. agency heads exhibited a much more thorough understanding and empirical analysis of the near impossibility of applying this procedure, of doubtful antecedents and no clear record of success anywhere, with a universe of peers each trained in a narrower disci-pline, to interdisciplinary enterprises which intrinsically could not be judged *as a whole* by *any* specialist peer.

   Alternative procedures to traditional peer-review are essential for evaluating all interactive research if one is to advance such activity. And some such were in fact described approvingly (see Recommendations). Among these were the use of "strong managers" experienced in the spe-cific interdisciplinary field, or various formulas. The simplest by far, where it applies, as for example to university-industry work is a formula for ***matching actual hard dollar support*** (always given after intense peer review by its own managers) provided by industry.

Recommendations*

1. Interdisciplinarity and interactive research (as defined herein) is much more important and significant and sure to increase faster than the academic culture recognizes; Universities must *plan* for this. *(ALL)*

2. The fact that the universities (in spite of major funding incentives by the government agencies supporting) them have been so resistant to innovation (for 40 years), clearly indicate that *new* approaches will be needed. Experimentation with really new structures (e.g. why have departments at all?) new *dis*incentives, new rewards *must* be undertaken. *(G.B., R.C., R.Y., H.Y, R.R.)*

3. Terminology must be clearly defined and used consistently across the community including the agencies, the National Academies, the University. Conflating multidisciplinary and interdisciplinary, interdisciplinary materials research with the academic departments of materials science and engineering etc. must be rigorously excluded. Interdisciplinary education's boundaries and structure need much more attention. *(A.L.B., G.B., R.C., R.R.)*

4. Research by sociologists and policy experts using industrial and international evaluations of interdisciplinary units in each nation's Universities is called for. This could determine the factors associated with success (*not* in generating good science but in performing genuinely interactive research) by a case-study of units regarded as successful by the outside world. (Self-evaluations of university programs by University personnel simply are not useful). *(A.L.B., R.Y., R.C., R.Y.)*

5. The *track record* of individuals and groups are a much more reliable indicator of future interdisciplinary interactions. The authorship of papers (whether joint or not) and the choice of journals (disciplinary or broad spectrum) and the nature of the approach (*requiring* collaboration

or not) and past record of collaboration, are objective data which can be used to immediately sift the genuinely interactive from the cosmetic. *(A.L.B., R.C., R.R.)*

6. The rapidly increasing collaboration with industry is the most successful part of I³R and its structuring can be used to bring about changes in universities. *(M.R., S.K., H.S., P.M.)*

7. Inter-national collaboration offers special opportunities, rewards, and effective interactions. A special case exists in the F.S.U. countries, especially Russia, where outstanding science can be salvaged by modest collaborative efforts from outside. *(Y.T.)*

8. Peer-review as institutionalized today in *most* agencies simply cannot be used successfully to evaluate interactive research. Research shows that both individual mail review and panel reviews mainly reflect and reward the work of those closest to the *reviewers. (R.B., R.R.)*

9. In its place, other approaches should be given much greater prominence in evaluating all interactive research. The DoD "strong-manager" has proved its worth. The use of past performance as the major factor, including the data as noted in Section 5 above is by far the most reliable predictor of success. Agencies have been too lenient and understanding when committing large amounts over long periods (40 years in half-a-dozen Universities) with meager results with regard to interdisciplinarity. Requiring proposing Universities to *show commitments* (e.g. by creating tenure granting interdisciplinary units, or tenured professorships in interdisciplinary fields) *before funding,* would really encourage a new *infrastructure of interdisciplinarity* so totally lacking today. The track record of success, by such objective metrics, should be a major factor in evaluation. *(A.L.B., R.B., R.R., R.Y.)*

10. Universities themselves should do vastly more introspection on their relatively poor performance (as judged by *their* peers in industry and

government) in *interactive research.* They should consider structural experimentation with departments and interdisciplinary units. At a minimum they could immediately institutionalize simple metrics and changes which could level the playing field much more for the obviously still struggling interdisciplinary fields—which are also their future, for example:

- Designating champions (*not* administrators) of I³R at the highest levels within the University, and identifying and cultivating faculty champions–practitioners of I³R.
- Stressing importance of I³R to students and faculty by appropriate awards and incentives.
- Counting all papers double when authors are from different departments, colleges, and institutions, for the usual academic bookkeeping purposes.
- Require evidence of interdisciplinarity for special, University ranks, chairs, alumni recognition, etc., including emeritus rank.
- Provide incentives for securing hard industrial dollars for research e.g. forgive part or all overhead; provide special equipment. (*A.L.B., R.C., G.B., R.B., R.R., R.Y.*)

11. Specifically for university-industry research, the following are worth considering:
    - Genuine, detailed, on site review by peers is performed routinely by industrial managers and personnel who size up University faculty research, before they make the decision to commit funds for such work. By far the most reliable, least bureaucratic way of encouraging such interactive research is for the government agency or Foundation to commit matching funds to the faculty member(s) funded by industry for research. These sums should not necessarily be dedicated *only* to the company's goals, but serve as incentive for more of such I³R with that company or others. The ratio of match

could vary from 1:1 to the 7:1 recommended by the Baruch Commission on Innovation under President Carter. *(R.R.)*

- With respect to the intellectual property posture of Universities in order to maximize interactive research (and bring in the maximum revenue from research) when corporations pay the full costs of the research:

1. All "intellectual property" (i.e. papers and patents) must reflect accurately the creators thereof whether faculty alone or including company personnel.

2. Patent applications if any should be filed at the company's expense in the names of the inventors and assigned to them.

3. An exclusive license should be available to the sponsoring company (if so desired). The terms of the license (to be negotiated by the inventors) could specify that (say) 25% of any accumulated revenue would be paid to the University. Thus the University achieves its goal of a revenue stream from major patents, without any bureaucracy. *(R.R.)*

* The initials following each recommendation refer to the principal originating champions, among others.

Preface ▸

"Interdisciplinarity" is emblazoned prominently across every national agency's proclaimed key strategies for funding modern research. Astonishingly, however, the intellectual effort put into analyzing, evaluating, or optimizing what is really meant by, or involved in, "interdisciplinary interaction" is, literally, vanishingly small.

The potential for serious problems for interdisciplinarity in the University world are obvious. The concept and reality of interdisciplinarity are squarely in conflict with the fundamental organizational principle of *all* Universities—which is to divide knowledge (quite rigidly for most purposes) into disciplines. Adjectives often used to describe the insularity of the disciplines are "silos" (see for example Heylin, in this report) or "walled-off from each other." Marcie Greenwood formerly the Deputy Director of OSTP in the White House used this metaphor of how Universities teach students baseball. They establish three disciplines of fielding, pitching and batting. These disciplines (departments) each establish faculties, undergraduate and graduate courses and do respectable (i.e. publishable) research, each in their own "disciplines." But they barely talk to each other and certainly no University institution exists to get them to play a game of baseball!! This is of course the price one pays for reductionism. Others have written in a similar context: "no world problem or opportunity comes in a discipline shaped box."

It is therefore an extraordinary omission on the part of both the academic community which claims to be doing it, and the agencies which claim to be supporting it, that so little open discussion and rigorous analysis exists on how this obvious goal of societally-relevant interdisciplinary research can EVER be done effectively in a discipline-dominated institution.

In the world of contemporary engineering and science, where the demands today to translate and transform knowledge to product much more rapidly, have led to a major compression of the time frame for every aspect of research, one aspect has suffered the most. That aspect is the thinking about the processes of research, itself. It is virtually unknown that a professional scientific/engineering society would hold sessions on the goals of the field, the *processes* of actually conducting the research, even on the best *processes* for research topic-selection or for distributing funding. (There are, of course, innumerable sessions on either the *status* of research funding, reports from agencies on what is being funded and how to *increase* funding, or enrollment).

This neglect of the "how" of research is not evenly distributed among the three major research players: universities; industry and government laboratories. Industry has indeed worked hard since the early nineties to completely overhaul its internal research structure. The dramatic elimination of untargeted (a-telestic) basic research from virtually every corporation worldwide, and outsourcing of much other research, is plain for all to see. Amazingly it has caused no re-thinking in the Universities or government. Among the long-term process changes in modern research, the one is that is more talked about in press releases, RFP's, and proposals, (no other parameter is even a close second) is the call for *interdisciplinarity*. One would then imagine that the study of "interdisciplinarity" would be a major focus among social scientists and science policy professionals. Yet, the literature on the topic is amazingly thin. This conference is a modest effort to change that situation.

The Universities
The historical fact is that "interdisciplinarity" was moved onto the 'screen' of science policy-makers in the late 1950's by industrial research leaders,

specifically William O. Baker, then V.P. Research of Bell Labs, and C. Guy Suits, V.P. Research of the General Electric Company, who recognized that the University world was not doing research nor training people in what had become imperative in industrial research: a totally interdisciplinary framework, specifically in materials research. Baker via his connections at the highest political levels was able to convince the Defense Department that they should explicitly try to shape university research into the inter-disciplinary mold mainly by offering unheard of *lengthy* (5 years) and *large* research contracts to those Universities that would conduct research in this new interdisciplinary mode. Thus was born in 1959 the DoD's Advanced Research Projects Agency's IDMRL (Interdisciplinary Materials Research Laboratory) program. Materials Research was the very first theme, which received this special attention.

As in all agencies, the setting up of IDMRL's was based on proposals by Universities which *promised* to do interdisciplinary research. Performance was another matter. Only two years later it was ARPA again which realized that not only was interaction required across disciplines *within* one University, but in order to make such costly knowledge valuable to the DoD customer, it was necessary to transition the knowledge across the institutional barriers *between* universities and industry. Another programmatic thrust set-up pairs of university and industry institutions, to experiment with this second direction of interactions. Thus *materials research* became and has remained by far the most significant example of programmatic national agency efforts to change the culture of the Universities from quintessentially disciplinary to genuinely interactive across disciplines, sectors, and institutions. Soon after ARPA, in the sixties, the AEC (now DOE) and NASA, also triggered the creation of interdisciplinary units on other campuses by providing large "block" grants. A total of some twenty such existed by 1972, when due to the Mansfield Amendment, the ARPA program was taken over by NSF, and (significantly) the adjective "interdisciplinary" was dropped from the title. Significantly also only a single university (Penn State) in the nation had institutionalized a MRL without any connection to any specific Federal contract.

In 1999, the track record of the 40 year long strategy, of offering financial *incentives* to Universities to enhance interdisciplinarity shows that it has had virtually no impact on the intrinsic problem. Universities remain the last bastions of narrow specialized insular subdivisions of knowledge. Interdisciplinary *projects* are strategies for—to put the best interpretation on it—encouraging cooperation and interaction in a set of 5-25 faculty from different disciplines. In every single case this is still conducted within the constraints of a discipline-dominated administrative and reward structure, which even in the best case must to greater or lesser degree inhibit the interdisciplinary interaction. Most importantly after the 5 or 10 year life of most such efforts, everything reverts back into the old disciplinary structure. This is a key flaw: *short term project support is useless for institutionalizing interdisciplinarity.*

Industry

This "unexamined life" of disciplinarity is fortunately only true of the University world. In industrial research the situation, especially today, is the exact opposite. All discipline-mimicking labs and titles have disappeared. Industrial research is structured by purpose and/or product. Virtually all research in major companies is team research. Virtually all teams are interdisciplinary, and interactive in all dimensions.

Government: Funding Agencies and Laboratories

The initial structuring (ca. 1950)of government agencies, unfortunately used the discipline structure of the Universities, because, the agency's job was seen as helping the University do its work. But as the mission of every agency has come to dominate funding categories, the structure has *slowly* moved more towards the industrial model of mission or task categorization.

Within government laboratories or mission agencies this change towards specific tasks or goals as the motif of administrative structure has gone even further.

Status Today

It is exactly 40 years since the U.S. Federal Government specifically intervened via one program in one agency (ARPA) to tilt the universities back towards interdisciplinarity. Has that goal changed or weakened? On the contrary this goal is now universal for at least part (sometimes the majority) of most agencies' funding. An order of magnitude estimate is that one billion dollars has been spent to advance the cause of interdisciplinary materials research.

What of the response by Universities? It is no doubt true that much very fine interdisciplinary research has been done. What is no less true however, is the fact *that there is not a single model, not a single example* of *permanent institutional change in any University, which is more favorable to continuing interdisciplinary interaction.* This is a startling result. It is therefore unanimously accepted on every campus that any interdisciplinary effort is a continuing uphill battle. At best then the agencies and foundations that support such interdisciplinary work accept the fact that it will be carried out in a hostile environment, without any complaint or demand for change on the part of the University or the agency.

At the same time the realities of the world—the necessity to *play* baseball—have obviously impacted every field, not only materials research. Every issue from the environment to medicine to every major technology absolutely demands an interdisciplinary approach. Indeed it demands more.

Interdisciplinarity Is Not Enough: Other Kinds of Interactions Are Needed: "I²R"

It has been noted that ARPA very shortly after starting IDMRL's also funded specific University-industry partnerships to learn how to do collaborative research across the university-industry interface. Today, while *interdisciplinarity* typically refers to the interaction across disciplines on one campus, such as: *Chemistry-Physics-Biology-Materials Science-Mechanical Engineering-Electrical Engineering*, in fact research in each of these departments may also be done cooperatively with an industry, or government laboratory, involving the same or different disciplines at the other institution. This is inter-institutional research.

Thirdly, the enormous pressure to move knowledge (even interdisciplinary knowledge) towards a product demands interaction along the chain of research-development-testing-evaluation, sales and marketing. This is inter-sectoral research. Finally, all three kinds of interaction may be involved simultaneously. A particular project may involve a University and industry, with team members in each drawn from various disciplines, and the goal may include very basic research as well product optimization.

Hence, all the kinds of research, which were discussed at this conference, could be most accurately described under the umbrella of the term *Interactive Research*. The logo developed for the conference captures this discussion rather well.

What today's scientist or engineer, or industrial manager, or funding agency or foundation must *necessarily* be concerned about, is how to most effectively carry out research which fits either on the binary edges or ternary area within the triaxial diagram.

The Program for the IDR Conference

The program was organized in the following sequence of sessions:

Session I: Present at the Creation

a. Historical Perspectives From Those Present at the Beginning

b. Panel 1: Driving Forces for/Barriers to, Interactive Research

c. Panel 2: Driving Forces for/Barriers to, Interdisciplinarity in Universities:
Teaching and Public Service

Session II: I3R: the Status Worldwide

a. Overview

b. Panel 3: The World Status of I³R

c. Panel 4: I³R: In Other Subject Areas

Session III: I³R Work: Overcoming Barriers: The Agents of Change Toward I³R

a. Panel 1: Role of Government

b. Panel 2: Role of Industry

c. Panel 3: Role of Professional Societies and Their Journals

d. Panel 4: Lessons Learned from Real-Life Examples of I3R Research

Session IV: Output to R & D Administrators and the World

a. The Interdisciplinarity Imperative

b. Re-inventing Interdisciplinarity

c. Panel 1: Overcoming Departmentalism Institutionalizing Interdisciplinarity

d. Panel 2: Smoothing the Interface: Univ.–Industry: University-Government;
Government-Industry Coupling

    e. Panel 3: Overcoming the Bias Toward 'Basic Science': Changing
       National Attitudes to Applications-Driven Basic Research

The presenters included key figures involved in the creation of interdisciplinarity in the field; senior university and industry managers who have had 25-40 years experience in the field; funding agency representatives, and very senior managers from government and industrial research labs from all major countries in the world. The detailed program is reproduced in the appendix.

The following report follows a modified sequence.

Acknowledgements ▸

The organizers gratefully acknowledge the financial support of the Alfred P. Sloan Foundation, which made the conference possible.

The National Science Foundation also aided the conference by providing a grant for travel support.

The Theory and Practice of "Interactive" Research (I³R)

The Interdisciplinarity Imperative to Create New Knowledge and Uses of Knowledge Across Boundaries of Disciplines and Institutions ▶▶

George Bugliarello
Former President, Polytechnic University of New York
Founding Editor of Technology and Society, Editor of NAE Bridge

A society in which knowledge has become the fundamental theme a theme as dominant in the practices and organizations of society as it is in science and technology cannot afford not to use any possible device, to adopt any possible paradigm that will enhance the generation and utilization of knowledge. Interdisciplinarity, which has already shown its significance in biochemistry, bioengineering or geo-chemistry, just to mention a few examples, creates new knowledge and uses of knowledge across boundaries of disciplines and institutions. But beyond being a powerful engine of creativity, interdisciplinarity is also an instrument of integration of the uses as well as of the users of knowledge.

Interdisciplinarity can provide new understandings, new horizons, new methods, new impacts, and new paradigms and lead to the creation of new disciplines. When very successful, the hope is that an interdisciplinary effort, rather than being merely evolutionary, adding incrementally to existing knowledge, will be truly revolutionary. An example is the development of molecular biology with the ever-widening impact that the discovery of DNA is having on knowledge, on technology and on society. Thus, if society is to draw maximum advantage from interdisciplinarity, it cannot afford to adopt a laissez-faire attitude, but needs to pursue interdisciplinarity aggressively and systematically. The hope that interdisciplinarity offers to the creation of new knowledge is immense, as opportunities can be found at the interfaces between virtually every set of disciplines and fostered by all sorts of multi-institutional or inter-institutional arrangements.

What is a discipline?

An overview of interdisciplinarity must start with an understanding of what a discipline is. A discipline is ultimately what people agree is a discipline. In other words, a discipline is a culture; a set of shared assumptions as to its foundations, scope and methods. In a very simplified way, one could look at a discipline as the result of a process that starts with a set of facts, which lead to hypotheses, which in turn lead to theories (Fig. 1). An ensemble of theories and explanations forms the discipline. The formation of a discipline often arises out of sub-disciplines, which eventually may become disciplines, as in the case of bioenergetics.

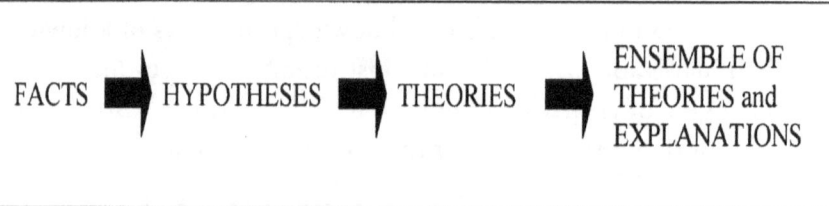

A simplistic historical morphology of the evolution of disciplines (Fig. 2) could start with a pre-disciplinary general knowledge which then expands along preferential directions, as was in the case of the Middle Ages with the trivium (grammar, rhetoric, logic) and quadrivium (algebra, music, geometry, and astronomy), which was deemed to encompass what was the important knowledge for the times. Those preferential directions could all be encompassed by an accomplished single person. Eventually, they became definite channels that, as knowledge increased, branched out and narrowed. Thus the channel that was chemistry branched out into inorganic and organic chemistry, or, the channel that was engineering branched out in the second half of the nineteenth century into civil, mechanical, electrical, etc. Periodically, new fundamental paradigms such as the theories of evolution, relativity, and quantum mechanics or, recently, deconstructionism, sweep across these channels and change the way we look at the world. This leads to new understandings, new methods (such as input-output analysis applied to economics), and new instruments (such as x-rays). In this context, we may observe that:

- Often, the branches of different shoots of knowledge come close to each other, as in the case of materials engineering and the physics of the solid state, or biochemistry and molecular biology.
- The more the ensemble of knowledge branches out, the greater is likely to be the danger of losing contact with a general view of knowledge.
- The ensemble of branches leaves many areas that are untouched and thus offers opportunity for further expansions.

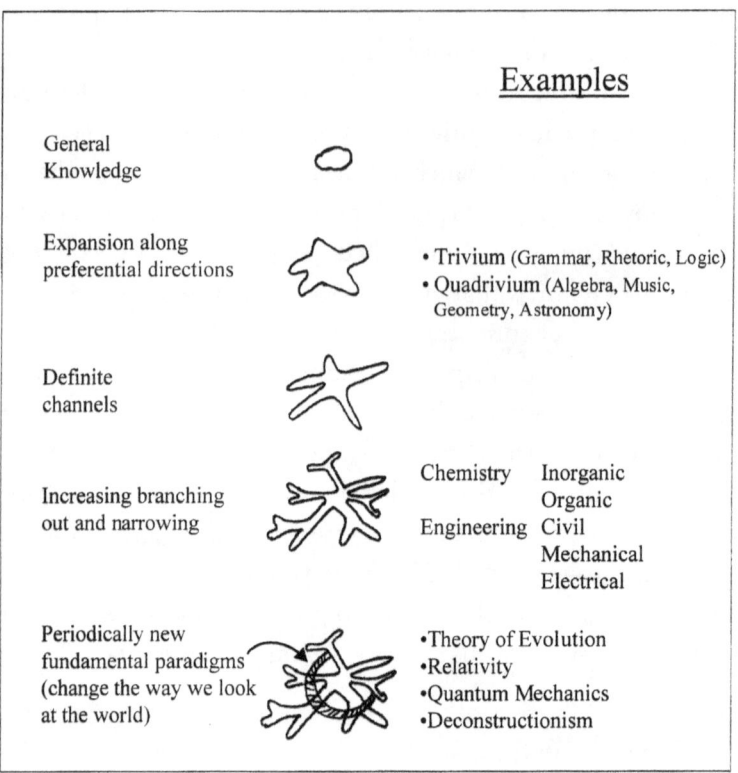

Fig. 2. A Simplistic Morphology of Evolution of Disciplines

As knowledge expands like an oil drop, the human capability to comprehend it, delimited by an arc on the oil drop circumference, cannot expand correspondingly and the sectors cultivated by a discipline become increasingly small fractions of the whole (Fig. 3). Hence the increasing importance of counterbalancing this effect by attempting to see the whole (integration) and establish connections across the drop (interdisciplinarity). In a constantly expanding domain of knowledge, the synthesis of immense knowledge domains is nearly impossible. But not so should be the hopes to harmonize basic tenets of major sub-domains, such as law and engineering, science and spirituality, technology and art, even if today we are far from doing so in most cases.

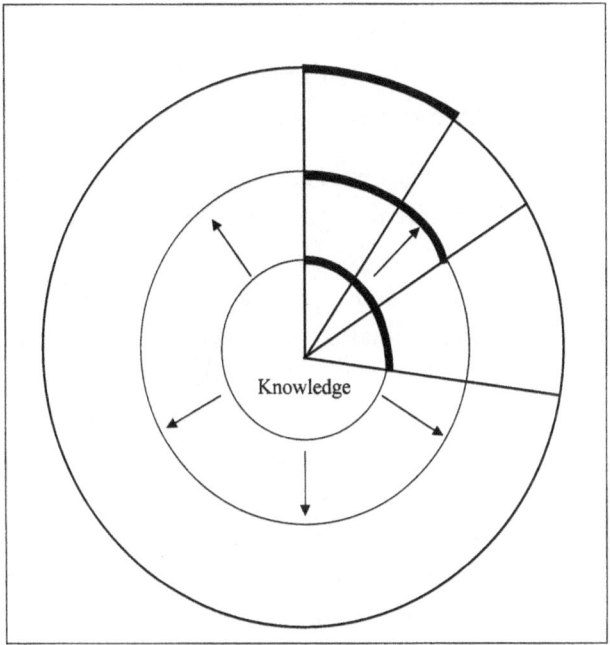

Fig. 3. The Expanding Oil Drop Analogy

Interdisciplinarity usually cannot be accommodated within a strong discipline-oriented culture such as ours still is. Thus, interdisciplinarity is not easy and is not for everybody. If only lip service is paid to it, it is a waste of energy and opportunities. However, not every potential interaction needs interdisciplinarity, that is, the *synthesis* of different disciplines. Alternatives may be an expansion of an existing discipline or multidisciplinarity, that is, *communication* between different disciplines, in principle a different and far less difficult task. Also, not all interdisciplinary opportunities have the same importance or potential. There are doors to treasures, but also to dead end alleys.

Maps of Interdisciplinarity

The development of interdisciplinarity can be enhanced by maps of knowledge, that is, by knowledge-driven paradigms and by institutional arrangements, that is, by organization-driven schemes.

There are many ways in which we can derive knowledge maps in order to identify opportunities for interdisciplinarity. We can consider, for instance, teleological aspects of knowledge, that is, aspects that lead to interdisciplinarity across different purposes of knowledge; or thematic aspects, leading to interdisciplinarity across different themes of knowledge; or what could be called domainial aspects, leading to interdisciplinarity across different domains. There are obvious overlaps among these different kinds of maps, depending on what aspects we emphasize as we attempt to classify knowledge. However, the first rule of interdisciplinarity is not to get hung up by these or other classifications because classifications are always arbitrary constructs of convenience that shift with the development of knowledge itself, and with social and historical perspectives.

Teleological Interdisciplinarity (Fig. 4)

The quest for knowledge can have a variety of purposes, such as:

- Explanatory: to explain phenomena and the world around us
- Predictive: to give us information for creating designs that will enable us to modify that world in a predictable way, tangibly or intangibly
- Normative: to give us the knowledge necessary for establishing rules that prescribe our actions
- Spiritual: to guide us in approaching the unexplainable
- Artistic: to guide the emotional expressions of our creativity
- Action-Oriented: knowledge, instinctive or otherwise, as a base for doing things not encompassed by the other categories in this classification

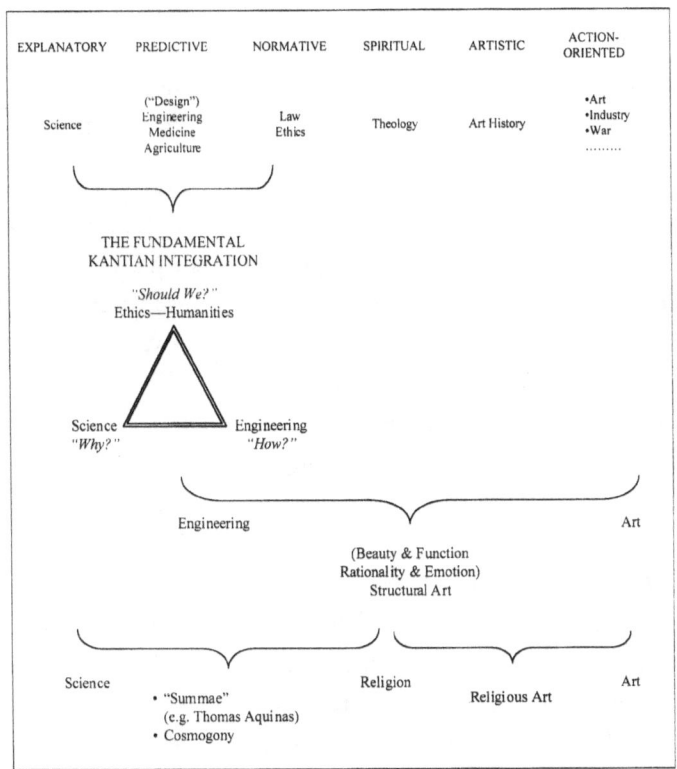

Fig. 4. Examples of Teleological Classifications
(the Purposes of Knowledge)

Explanatory knowledge encompasses, obviously, science. Predictive knowledge encompasses design, broadly intended, that is, disciplines such as engineering, medicine and agriculture that have the purpose of modifying nature (as in the case of a physician fighting that natural phenomenon that is disease, or of an engineer creating a device or changing the course of a river). Normative knowledge encompasses the law and ethics, and spiritual knowledge, theology. An artistic discipline is art history, whereas we could classify art as an active or action-oriented endeavor. To reiterate, however, these categorizations, if they are to be helpful, cannot be rigid.

The interaction of explanatory, predictive and normative aspects of knowl-edge is exemplified by the fundamental integration asked for by Kant of the *whys* (science), the *hows* (or, as I have broadly defined them, engineering) and the *should we* (that is, the ethic-humanistic aspects of knowledge).

The integration of the three Kantian questions can be represented by a tri-angle with the three questions as vertices. The integration of engineering and art is one of function and beauty, or rationality and emotion; it has led, for instance, to structural art, but it is still embryonic at best. The integration of science and religion has led in the West to the medieval attempts of the Summae, such as the heroic one of Thomas Aquinas, who tried to deal with seemingly every more irreconcilable set of domains. In another direction, the integration of religion and art, that is, of the spiri-tual with a facet of action-orientation, gave rise to religious art.

Domainial Classifications (Fig. 5)
The domains of knowledge are virtually infinite. Among the broadest examples of domainial classification is one that looks at the domains of biological organisms (that is, the domain of biology), of machines (that is, the domain of engineering), and society (that is, the domain of human interactions). The synthesis of these three domains what I have called, for short, the *biosoma* synthesis can, again, be represented by a triangle. Examples of interdisciplinarity in this context are socio-biology at the interface of biology and society, bioengineering at the interface of biology and machines, and socio-technology at the interface between the social and the technological disciplines.

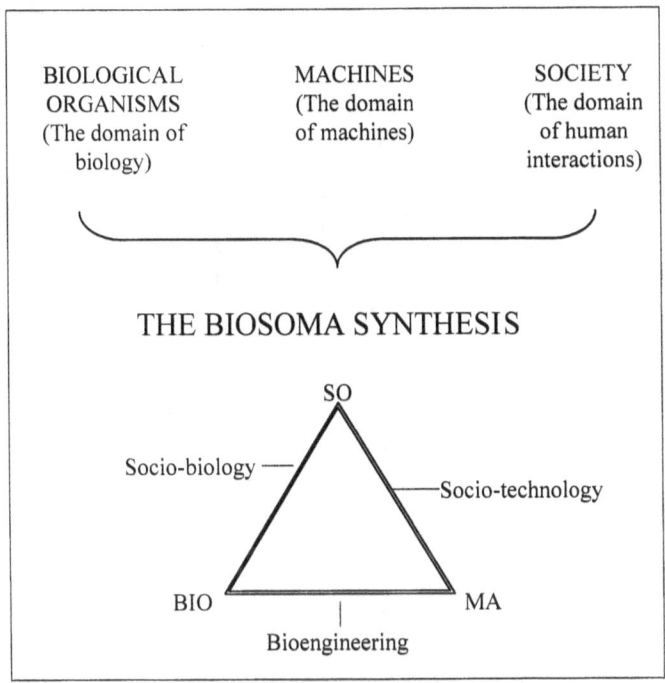

Fig. 5. A Domanial Classification
and Examples of Biosoma Syntheses

Kantian and biosoma integrations have been pursued for a long time in various ways, many of them ante literam, all the way to today's concept of consilience by E.O. Wilson (Fig. 6). The litmus test of the completeness of the two integrations is whether their respective triangles are closed. It is clear that, thus far, they are not, except perhaps for some current embryonic attempts at science-technology-society syntheses and at biosoma synthesis.

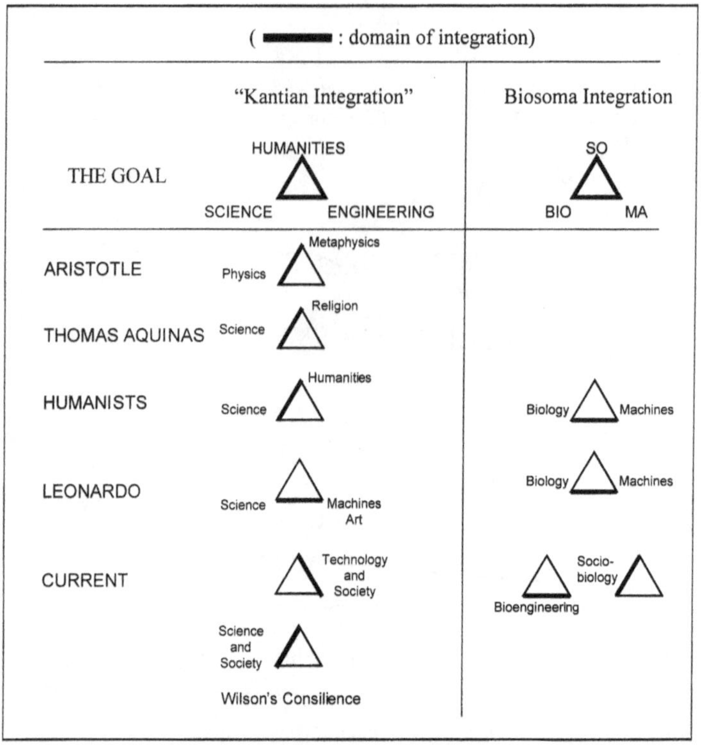

Fig. 6. The Quest For Integration: Historical Examples of
Kantian and Biosoma Integrations

Thematic Classifications (Fig. 7)

Thematic classifications look at knowledge in terms of specific cross-cutting elements or themes, such as materials, energy, information and systems, or dimensions, from pico to macro, or forces, or speed (such as slow- versus high-speed fluid mechanics), or performance (for instance, from entities of definite performance, such as a well designed machine, to elements of semi-definite performance, such as a social entity or a biological organisms, neither of them totally predictable in its performance, to elements of indefinite performance, such as a work of art, that cannot be created with any certainty that it will have the desired effect on the persons viewing or experiencing it).

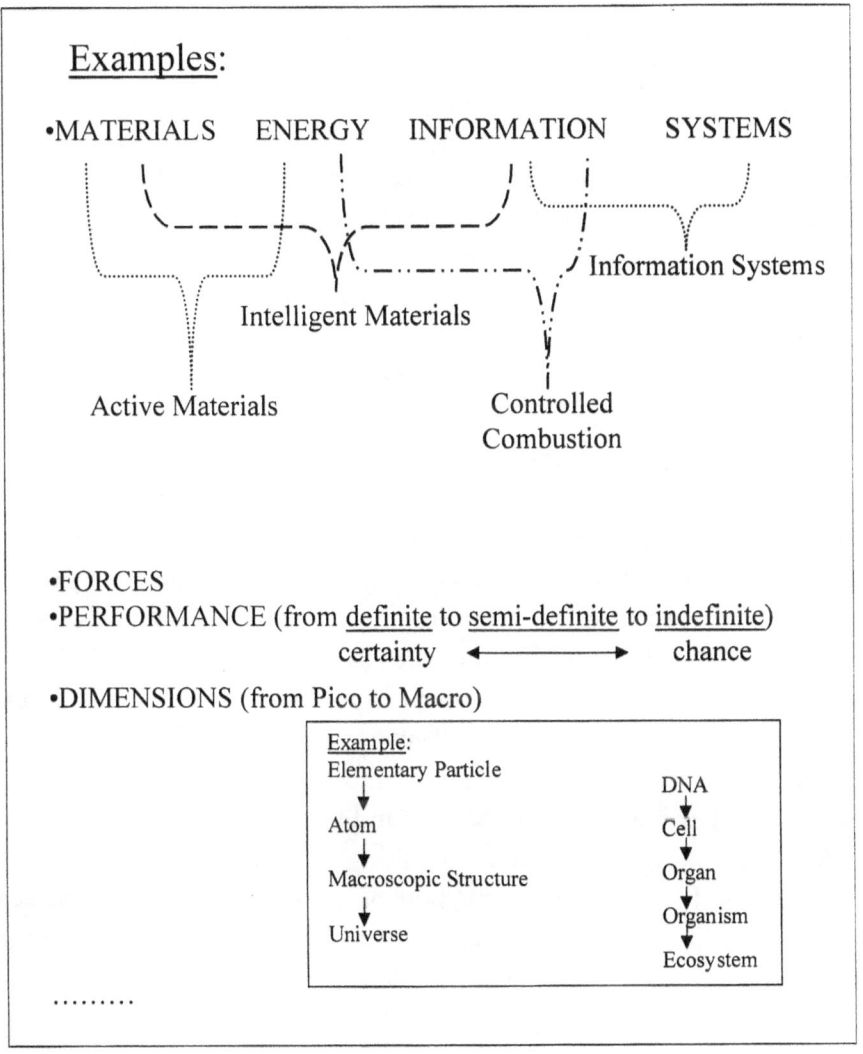

Fig. 7. Thematic Classifications

The interaction of concepts of materials with concepts of information has led to the integrated domain of intelligent materials; of concepts information and systems, to that of information systems; of concepts of materials and energy, to the domain of active materials; and of concepts of energy and information, to the domain of controlled energy processes, such as controlled combustion.

In terms of dimensions, interactions and integration may occur at various levels, from pico to cosmic. For instance, at one of the smallest levels of a biological organism or of an instrument, say level 1, structures, that is, materials, may be integrated with elements collecting and elaborating information and with energy activators. In turn, this first level may become part of a larger system, call it level 2, which may become part of an ever larger system (level 3), and so on (Fig. 8). The integration difficulties vary with the size the scale of the domain. It is a matter of speculation and viewpoint whether in biology and sociology they tend to increase as the size of the domain increases, it being possibly easier to integrate small organisms and organizations than an ensemble of them. But even at the smallest size level it is an enormously difficult task if one considers the complexity of the single cell or, at an even lower level of dimensions, of DNA, or the complexity of a single individual as the base element of a social group. In the case of machines, the difficulties may be greater at the first levels, if one considers, for instance, nanoscale self-replicating devices; somewhat less at intermediate levels, such as that of an automobile, where one integrates generally well understood components; and very large again in the case of devices such as a space station, requiring the complex integration of machines and biological and social systems. Different purposes are served by integration at different levels. Thus, integration of a pacemaker with the organisms it serves by no means a trivial task serves a different purpose than the large scale integration of an ecological system.

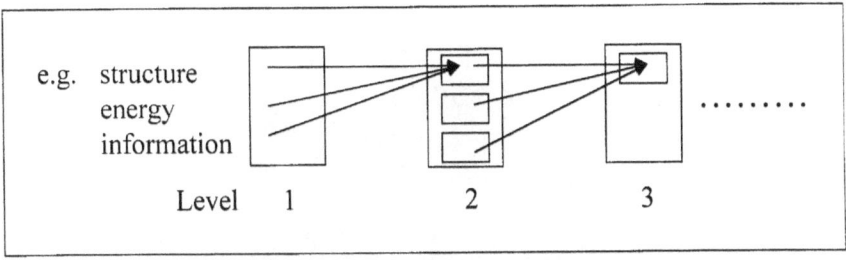

Fig. 8. Multiple Levels of Integration in Terms of Dimensions

Combinations of Multiple Kinds of Mapping

Combinations of multiple kinds of knowledge maps offer an enormous panorama of interdisciplinary possibilities. For instance, integration of the dimensional and domainial aspects of knowledge may include integration across the physical and biological domains, such as physics at the atomic level and DNA, or of molecular chemistry and studies of the origin of life, or of science at the nano and pico levels with nano and pico engineering.

A simple but powerful paradigm for combining teleological, domainial and thematic interdisciplinarity, such as that in Fig. 9, shows the possible interaction of science, engineering and the humanities with biology, society and machines, and with materials, energy, information and systems. Interdisciplinarity can occur in any direction of the diagram, as the examples at the bottom suggest. However, interdisciplinarity can also occur within each block of the matrix. Thus, in the materials block, interdisciplinarity occurs among disciplines dealing with metals, ceramics, polymers or composites, and in the humanities block, interdisciplinarity may occur among ethics, other philosophical areas, history and literature.

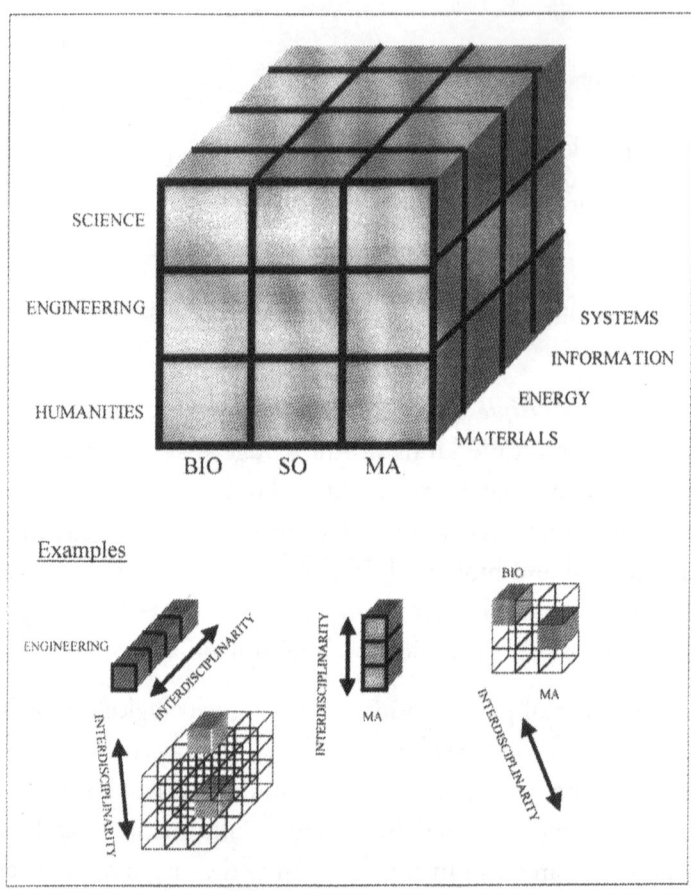

Fig. 9. Example of Paradigm for
Combined Teleological, Domanial and Thematic Interdisciplinarity

The diagram in Fig. 9 is but one of the many ways of exploring interdisciplinary possibilities. Clearly, there are many other possible paradigms for identifying those opportunities. In searching for them, serendipity should not be discouraged by utilitarian or short-range pressures, even if it were to lead to the exploration of interdisciplinarities between seemingly very distant areas such as literature and materials. Mainly, however, the quest for interdisciplinarity stems from a compelling need, as manifest in the interaction of engineering and medicine or in the concerns spanning science and society.

Institutional Arrangements

The interdisciplinary imperative is also an institutional imperative. As an interdisciplinarity effort develops, it needs organization and support. The chances of success increase with the flexibility of the environment in which interdisciplinarity is being pursued. In this, there is clearly a the progression of difficulties. The easiest environment should be within a department and the difficulties tend to increase when interdisciplinarity requires collaboration between departments, between colleges, between similar institutions, or between different institutions, such as a university and a national laboratory. However, this scale of increasing difficulties is not always monotonic, as it depends very widely on local circumstances. Thus departments are not always the most favorable environments.

Institutions have responsibilities with regard to interdisciplinarity, if they accept the fact that interdisciplinarity is a powerful instrument for the discovery of new knowledge and for the integration of knowledge. Institutions need to encourage interdisciplinarity by providing the interdisciplinary efforts with adequate resources, by having them report to hierarchical levels that are fair to them and understand their challenges, by promulgating proper promotion and tenure policies, by endeavoring to integrate the interdisciplinary efforts in the fabric of institutions and, ultimately, by encouraging the rest of the institution to respect the difficult efforts that interdisciplinarity represents.

An institution that believes in the value of interdisciplinarity needs to provide or support the leadership of interdisciplinary efforts and also provide leadership for the administrative support of such effort. At the same time, however, it needs to control the process of the generation of the efforts by establishing criteria for the allocation of resources. (It could, for instance, follow the approach used by venture capital in new enterprises by taking risks in accepting that some efforts will fail, but that such failures will be balanced by greater success of some other efforts). The institution also needs to develop a process to evaluate the efforts and provide oversight a process in which it should involve also outside advisors.

Institutional arrangements cannot overlook the fact that sociological factors are important. Traditionally, reputation and advancement come more easily from specializing in an established discipline in a narrow field. Yet, ultimately, pioneers in a new discipline may receive more recognition and, more importantly, contribute more to the advancement of knowledge than the plowing of a well established field. The collaboration across different disciplines is not always easy. Consider, for instance, the difficulties between medicine and engineering whereby, to this day, the medical doctor does not usually consider the engineer his peer. There are clearly questions of pecking order and also of reward, that an institution, and the whole scholarly community, must consider.

Evaluation of Interdisciplinarity

Different viewpoints, from the institutional to the national to scholarly one, need to be considered in evaluating an interdisciplinary effort. The key question is always: what should one ask? By way of example, from the scholarly viewpoint, one could ask:

- Where does the interdisciplinary effort that is being evaluated fit on a map of knowledge?
- Is the effort truly interdisciplinary or is it multidisciplinary? As a corollary, is it departmental, multidepartmental, or interdepartmental?
- What new insights have been gained? Have new syntheses been achieved?
- Is it the basis for a new discipline?
- What are the impacts, actual and potential? Educationally (is it teachable?); on research (views, methods, instruments?); on other disci-

plines (does the effort enhance them or make them less relevant?); on society (on the economy, on the community, etc?).

- What is the panorama of probable or possible future developments?
- What is the effectiveness of institutional arrangements?

Although today we still observe stove pipes in many institutions, and intercultural or interdisciplinary communication still comes hard, we need to recognize that there are important success stories. Centers of excellence and a wide variety of other interdisciplinary centers and institutes have multiplied and new disciplines have emerged to make just a relatively new example, environmental science and engineering.

The interdisciplinarity of entities, broadly intended, includes the interactions university-industry-government-community that lead to the creation of those new instruments of opportunity for a knowledge society that can be called knowledge parks a term that can be used to encompass science parks, university-industry parks, science cities, etc.). The record of universities and other knowledge institutions in creating these new instruments is quite spotty, often because of inadequate outreach to communities or inadequate commitment by the institution or the community. Yet, Silicon Valley, or the Research Triangle Park, or Metrotech at Polytechnic University are among several very clear examples of the potential and impact of knowledge parks.

What Does This Mean for Materials Science and Engineering?

In the context of a gathering promoted by materials scientists and engineers, it is appropriate to ask: what does all this mean to these fields? From earth dynamics and biological tissues, to material applications in energy and information, all the way to complex systems like a space station, it is obvious that the chain that links, in the field of materials, natural phenomena to the human-made offers an enormous scope for interdisciplinarity. Interdisciplinarity is needed to forge new concepts and to establish connections between the elements of the chain, as well as between the chain and societal needs. The continuously evolving relation between what society needs and what materials could do to satisfy the needs will place ever greater demands on interdisciplinarity, as essential to achieving goals such as self-assembly and self-repair, or ever more extreme performance in terms of strength, conductivity, and information-holding capacity, or more favorable processing characteristics (such as manufacturability and recyclability), or desirable use characteristics of materials such as ease of employ and methods of joining. The greatest interdisciplinary challenge is whether advances in these domains can be truly revolutionary rather than evolutionary.

The need to respond to these challenges, however, should not make materials scientists and engineers overlook the importance of their interdisciplinary involvement in the great questions of nature, ranging from those directly relevant to materials, such as the cosmogony and the biogony of materials (that is, the creation of materials by living systems, with the associated question, is biogony occurring anywhere else in the cosmos?), to the very fundamental questions of the very nature of nature that stretch from before Pythagoras to the theory of strings. Is it mathematical? Is it material? Are there laws? Is chaos the reality? Or teleology? The field of materials, like any other major field of knowledge, needs to address these questions from its perspective, because only out of an ensemble of perspectives can new syntheses emerge that will help us better understand nature and our role in it.

Policy Making and Peer Review in the UK Engineering & Physical Sciences Research Council ▶▶

Richard Brook
Engineering and Physical Sciences Research Council
United Kingdom

Introduction

One of the recommendations in the UK Government White Paper "Realizing Our Potential," published in 1993, was that the structure of the research councils should be changed. The setting in place of the Councils (April 1994) was recognized as a special opportunity to bring about major changes in the pattern of research council operation. The new Councils had no obligation to take over the structures which had been present in the former system. Within the EPSRC, the opportunity was taken to introduce new methods for the setting of research priorities and new methods for the selection of particular research projects matched to those priorities. In the following pages, a description is given of the operation of the new system; it indicates something of the reasons for the changes that have been made, it describes the mechanisms that are involved in the new procedures and it summarizes initial experience with the system.

Emphasis is given to two topics, namely, to policy formation where the role of the "generous generalist" has been a crucial element and to project selection where a system based on colleges of peer reviewers has been in operation. The opportunity which peer review colleges provide for constructive interaction with the overall process of project selection is also explored.

Policy Formation
The key role in policy formation, within the former SERC, was played by boards and committees formed by colleagues drawn from academy, from industry and from other sectors. This system involved a considerable hierarchy, as indicated in Fig. 1, there were some 750 people involved in this system at the time of the SERC's termination.

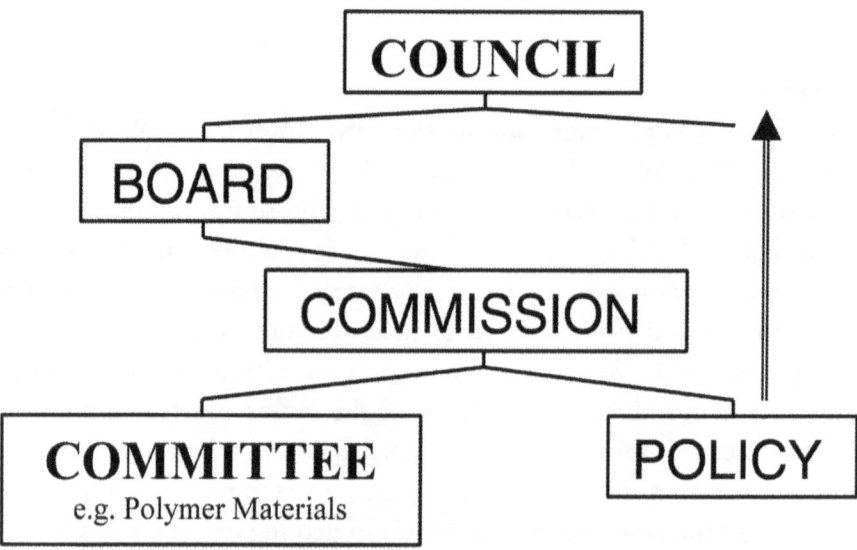

Figure 1: Policy making within the former SERC

Policy within the former system was initiated at the individual subject Committee level. Thus, for example, a Committee concerned with Polymer Materials would advise on required initiatives in this subject area. The advice would be passed (and often re-passed) to the Materials Commission, which would in turn pass it to the Engineering Board for transmission into

Council itself. The integrated policy proposals from SERC would then be received by the Advisory Board for the Research Councils prior to decisions being made on the eventual allocation of finance. The finance, one allocated, would then return down the hierarchical system.

This arrangement had the great benefit that many participants in the research community were active in formal policy formation. It had, however, a number of serious disadvantages. The first among these was the slow response of the system. Notification of decisions on requests for initiatives could take up to two years, a delay which resulted in a given Committee having to decide on policy at a time when it had not yet heard the response on the previous year's requests. This delay was a cause of considerable confusion in that choices had to be made between repetition of the previous year's policy (to achieve consistency) or identification of a new set of priorities (to demonstrate topicality).

A second problem with the former system was that the proposals lost precision as they mounted the hierarchy. Much of the detail was necessarily lost as proposals for initiatives were bonded onto other proposals and as successive Committees introduced modifications. A third problem was that the Committee programs themselves tended to become overly restrictive as a consequence of the direct involvement of the subject Committees in policy making. To take an example, the six Committees of the Materials Commission each had strategies (each of some ten pages) which were intended to act as guidelines to the proposing community. The presence of some sixty pages of strategic guidance can be seen as a considerable inhibition on the individual creative contributions, which are, after all, sought form those preparing proposals.

A major change introduced by the EPSRC has, therefore, been the departure from the Committee system. The organization of the EPSRC is as indicated in Fig. 2 associated with a division between policy formation and project selection. These two functions which resided together in the former subject Committees have now been separated and policy making is now conducted on an annual cycle in a manner distinct from that of individual project selection. As indicated in the Figure, policy is now presented in the form of a published set of program descriptions, the so called "Landscapes" where Council indicates briefly the pattern of research portfolio which it would wish to construct in each of its eight program areas (Table 1). The advantages of such a system are that it can be conducted on an annual cycle; indeed the policy forming exercise lasts some six months. It can also be focused, since it is not subject to compromise or modification resulting from discussion in a sequence of policy forming bodies. It can also be open in that the results of the policy thinking can be published and considered by all those intending to put proposals to the EPSRC.

Table 1: Program areas in the EPSRC

Science Programs	Technology Programs	Engineering Programs
Mathematics	Materials	General Engineering
Physics	Information Technology, and Computer Science	Engineering for Infrastructure, The Environment, and Healthcare
Chemistry		Engineering for Manufacturing

In developing its policy, Council takes advice and guidance from three sources. There is specific attention to the reports of the Government Foresight exercise. Advice is, in addition, received from two Panels of Council, the Technical Opportunities Panel (TOP) and the Users Panel (UP). These two Panels are, in the simplest terms, representative of the research provider communities and of the research user communities. The Technical Opportunities Panel, which is predominately formed from academic colleagues, has the role of advising Council on changes in the character of research disciplines themselves, i.e. on instances where a subject is showing rapid flowering or on instances where a subject is becoming exhausted. There is also the task of indicating areas of interdisciplinary research, which may not have been identified by specific communities. The User Panel which is formed predominantly from industrial colleagues, has the task of identifying those research and training requirements which can be identified as a consequence of the anticipated need of the industrial sector in the medium to long term.

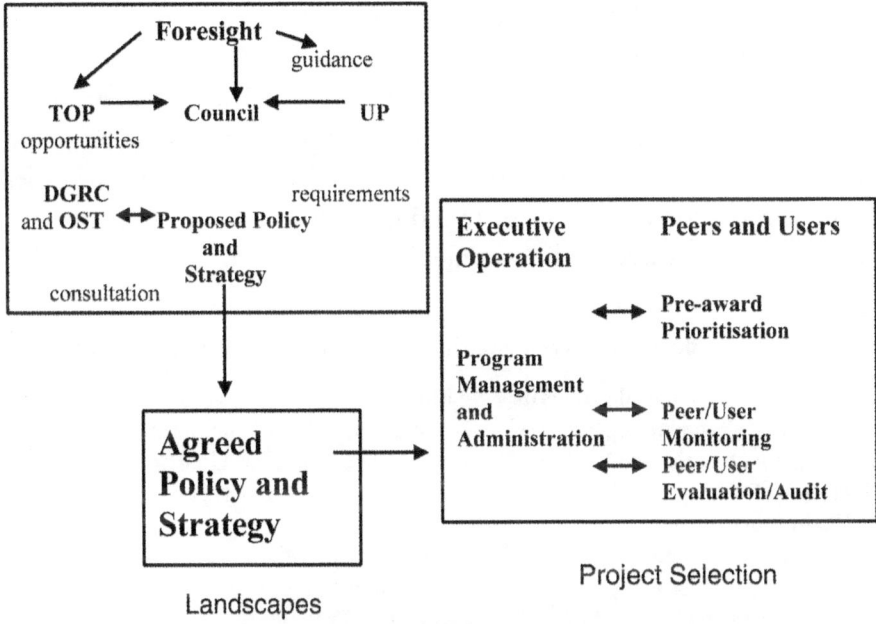

Figure 2: Policy making within the EPSRC

It may be helpful to report on the operation of the two Panels, TOP and UP. The first step in the sequence is that business plans are prepared by each of the program areas, the plans being then submitted to the individual Panel members. A business plan typically consists of a brief summary of the existing research portfolio in that program area, together with

comments on the program by professional institutions and other interested parties. The document also contains a summary of decisions made
by the Foresight Panels relevant to the program area. It finally contains a
statement indicating what changes would result in the program were
financing to be held at the same value, increased (by 10%) or reduced (by
5%) over the next three years.

These summaries of the research portfolio in each of the program areas,
are considered by the individual Panel members in the light of seventeen
criteria (Table 2). These criteria can broadly be divided into five groupings, namely: a set relating to the commercial or industrial promise of the
research: a set relating to the intrinsic intellectual interest of the research; a
set describing the ability of UK industry to exploit any results emerging
from the research; a set describing the ability of the UK research provider
community to conduct the work and, finally, set reviewing alternatives for
the support of research in the area under consideration.

Table 2: Criteria for determining allocations of finance to program areas

Potential Socio-economic Benefits	Ability to Capture Benefits
1. Economic Competitiveness	10. Strength of User Community
2. Provision of Basic Needs	11. Uptake Capacity of User Community
3. Physical Security and Safety	12. Potential for Rapid Technology Transfer
4. Health Improvement	
5. Skills Requirements	**Provider Capability**
	13. Strength of Provider
Nature of Research	14. State of Provider Base
6. Alignment to Foresight	
7. Research Potential	**Funding Considerations**
8. Pervasiveness	15. Importance of EPSRC Funding
9. Interdisciplinary	16. Funding Leverage Potential
	17. Capacity to Absorb Increase in Funding

The Panel members judge each program area in the light of the seventeen
criteria and advise colleagues at the EPSRC of their findings. At the Panel

meeting itself, each of the program areas is considered in turn. Panel members have available an analysis of their assembled viewpoint presented in terms of the strength of the area under consideration and against the given criteria, together with an indication of the standard deviation of the opinions represented by Panel members. The discussion of each Program area can then be conducted with full awareness of those points on which the Panel is agreed and of those points where differences of opinion arise.

At the close of the meeting, following discussion of each of the fourteen program areas, the Panel members indicate in secret vote those program areas which should receive greater support, those which can satisfactorily survive with lesser support and those which Program areas should be held at constant funding. The conclusions for the combined Panels in the 1997 exercise are shown in Fig. 3.

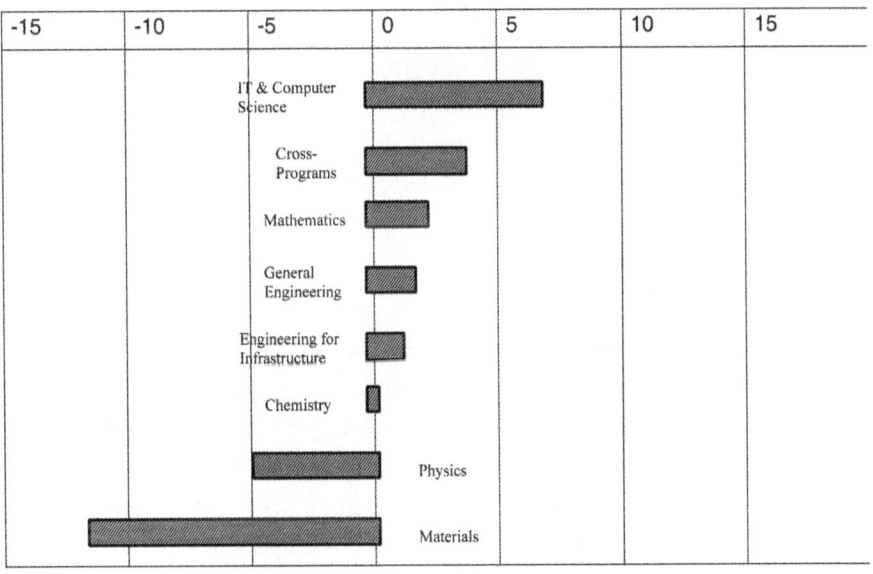

**Figure 3: Recommendations for change
in the funding allocations to program areas (1997)**

he advice from the Panels is translated into recommendations for changes in the financing of the individual program areas. This advice is then taken by Council and set against the actual financial resources in drawing up decisions on changes in program support (Table 3). As indicated, the entire exercise occupies some six months; it can therefore be satisfactorily fitted into an annual cycle. In addition to the indication of financial resources, the final policy statement from Council consists of the series of two-page landscape statements indicating those themes where, within a given program area, research is to be preferred.

Table 3: Commitment figures (sums available for initiating research projects) 1998/9

	Change from former plan for 1998/9
Increased	
Advanced Fellowships	25 → 32 Fellowships
ITCS Program	+ 3%
Mathematics Program	+ 10%
Engineering Taught Courses	+ 20 Studentships
Reduced	
Materials Program	
Physics Program	- 5%
	- 3%

To summarize the procedures for policy formation, the EPSRC now has a sequence which can be comfortably followed within a one year cycle and which has shown itself to be capable of setting priorities and of modifying support in the light of these priorities. The system is capable of paying close attention to the statements emerging from the Government Foresight exercise; it is also sensitive to the recognition of new research areas and of industrial requirements.

The two Panels can be seen as a substantial conceptual change in operation. The use of some dozen "generous generalist" Panel members in place of the determined advocates of specific disciplines, characteristic of the former system, has allowed a degree of cohesion to be developed across the full remit of the EPSRC research portfolio.

One difficulty, which has been identified in the operation lies in the preparation of the business plans for the Panels. It has, for example, been noted by certain of the professional institutes that the role of the business plan in influencing the thinking of the two Panels may be considerable. There is, therefore, a strong wish on the part of such institutes to become involved in the preparation of the business plan statements. While clearly contributing to the transparency of policy formation within the EPSRC, this step has been seen by the Panels as pushing them back to the "bunker mentality" associated with the former policy formation system in which defenders of particular subject areas compete for resource. The current view is that great advantage is to be gained by retaining the spirit of discussion within a generalist group; it is important to sustain procedures whereby an open pattern of operation can be achieved while allowing Panels to function in the manner intended.

Project Selection

In developing procedures for the building of a research portfolio matched to the policy put forward by Council, the EPSRC has throughout sought to make peer review the central element in the system. The overriding principle is that the best portfolio of projects will emerge if scientists and engineers are able to prepare research proposals against the brief but clear indications of the landscapes document and if they research proposals judged by a peer review system in which they can have confidence. To this end, attention has been given to the construction of a system in which the research proposer can play a strong role and in which the research proposer can play a strong role and in which a high degree of transparency can be achieved. The method adopted for the responsive mode has been based upon the concept of Colleges of reviewers.

In setting up this system, a first task has been to involve the research proposers in the selection of teams of reviewers. To this end, all those who had submitted proposals, successful or not, to the SERC in the range of subject areas now covered by the EPSRC were asked to put forward three names of colleagues (no more than two from the academic sector) whom they would trust to act as peer reviewers. On the basis of the nominations, and with due attention to the need for a comprehensive coverage within a given program area, College of referees were then established. Such Colleges typically now consist of 70 members, each of whom has agreed to review some 10 proposals per year in a prompt and sympathetic manner. The College membership is published. With peer review itself being conducted by the College system, the EPSRC has been able to entrust the operation of this system to Program Managers. Such managers have the role of receiving proposals and of then ensuring that they are fairly and effectively considered by the peer review system. The system itself is indicated in Fig. 4.

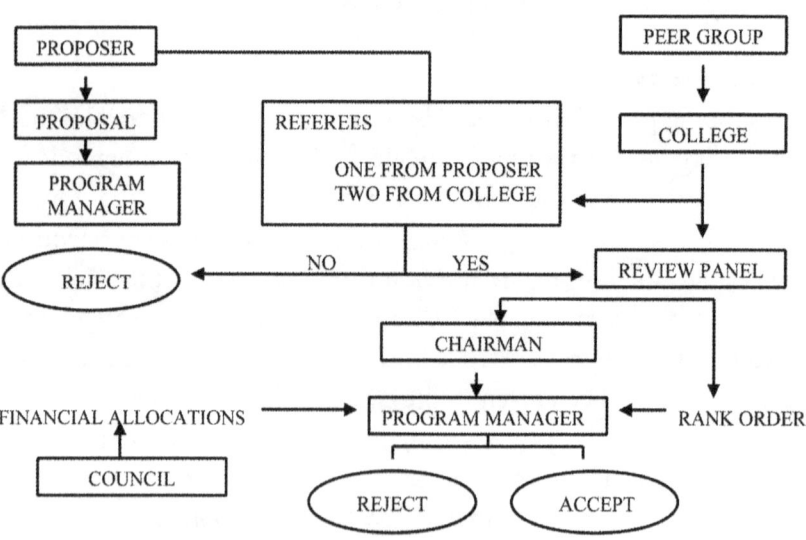

Figure 4: Flow sheet for peer review in the EPSRC

The initial reaction to this system has in some respects been favorable, notably in respect to the establishment of Colleges: the major disquiet has arisen in respect to the Program Manager role. A clear difference of opinion can be identified in respect to the desirable qualities of a Program Manager. Some colleagues in the academic community are by reference to the system of the National Science Foundation in the United States, persuaded that the role should be performed by a professional research worker, knowledgeable in the subject area and drawn from the relevant research community. As alternative viewpoint also well represented in the academic community is that the Program Manager should have no professional understanding of the research area in question; it is only under such circumstances, it is claimed, that the Program Manager will be disinclined to interfere directly in the peer review process. In general, the EPSRC has sought to appoint Program Managers who are familiar with the language of a discipline and who, more widely, are experienced in the research process. There has, however, been a wish to avoid the risks associated with the American pattern, namely those arising from the involvement of Program Managers in value judgments on the quality of research proposals. As noted earlier this is seen as the specific preserve of the peer review College members.

Two aspects of the new pattern of operation may be noted. The first is the consequence of having removed the deadlines for the submission of responsive mode grants. This change has resulted in EPSRC's receiving some 25% fewer proposals than were put to the former SERC system in the relevant subject areas. A number of reasons have been advanced to account for this reduction. The most persuasive is perhaps that of the removal of local pressures on applicants by university colleagues for the sending of proposals during the 'grant writing season'. The removal of the deadline has enabled academic colleagues to delay sending proposals until the point is reached where the proposal is itself fully finished and ready. In this connection, peer reviewers do report that the quality of proposals has been improved by the change. The second aspect relates to the role of Program Managers. Here, the task continues of fully persuading the community that they are sympathetic colleagues rather than opinionated judges. Individual Program Managers have been attempting to address this issue by way of town meeting, College gatherings and Newsletters. Although a lingering regret at the removal of the committee system is to be recognized, it is fair to say that the more overt forms of criticism made of the Program Manager system have declined, both in intensity and frequency.

Refinement of Peer Review

The previous section has indicated how peer-reviews operate within the IPSRC. It has however, given emphasis to passive aspects of the review process. The results are those which have emerged from allowing the research community to select approved colleges of reviewers and then of allowing these colleges to select approved research projects. It then becomes pertinent to ask if a research council is ever justified in interfering in this overall process.

The seriousness of this question is indicated by Figure 5 where the characteristics of the current peer review colleges are shown. The figure shows, for the six subject areas covered by the EPSRC, the average age of college members and the fraction of the college membership which is made up by women. It then becomes interesting to explore the characteristics of the research community which has selected this particular character in the colleges. It can then be found that the average age of principal investigators supported by EPSRC grants is 45 for Chemistry, 47 for Physics, 45 for Mathematics, 47 for Materials, 45 for ITCS and 47 for Engineering. The fraction of women principal investigators in EPSRC research projects is 6.5%.

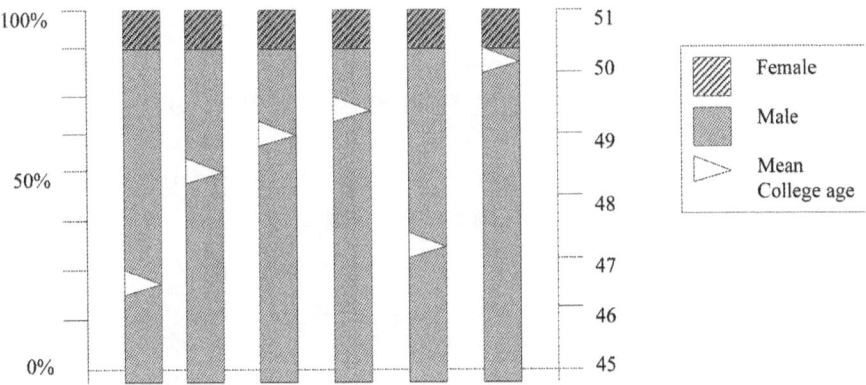

Figure 5: Gender balance and mean age of colleges

A cursory comparison would suggest that when the research community is asked to nominate peer reviewers, it chooses people who are remarkably like itself. There is furthermore the suggestion that when peer reviewers are asked to select principal investigators for research projects, they choose people who are remarkably like themselves. There is no intimation of questionable or improper conduct in this matter. It is simply a reflection of a strong feed-back loop and of the inherently conservative element within the peer review process.

If a research council then seeks a way of breaking into this cycle, then it is clear that the only opportunity for such deliberate design lies in the selection of the peer-review college. It would be a betrayal of the peer review process were the Council to interfere directly in the selection of the projects. A hypothesis, therefore, is that any one wishing to determine the future characteristics of the research community should seek to build those characteristics into the current peer review community. This is clearly a matter where great caution must be exercised but the EPSRC has taken a tentative step in making supplementary nominations to the colleges, first, of scientist and engineers who are under 40 years of age, and secondly, of scientists and engineers who are women.

The level of interference is small in that the additional complement is some 10% of the total. The mechanism is nonetheless seen as a way in which a gently re-shaping of the research community can be made with a view to ensuring a better participation in age and gender balance. It is, of course, tempting then to ask whether further refinements can be made in directions such as multidisciplinarity.

Conclusion

In responding to the requirements set upon it by the White Paper, the EPSRC has made substantial modifications in the procedures which are used both to establish policy and to select individual projects. The intent throughout has been to draw to the attention of the proposing community broadly defined but clear indications of sectors in which research activity may be expected to show promise. In meeting this overall objective, a number of assumptions have been made.

A first is that in selecting research projects the central mechanism must be that of peer review. To this end the College system has been established and every effort has been made to ensure the participation of the Science and Engineering community in a process which can be well understood and in which people can place confidence. A second assumption is that the creativity of the individual scientist or engineer must be the foundation on which research success is to be built. To this end, an increased emphasis has been placed on the responsive mode and on the refinement of this mode so that research objectives can be identified without in any way inhibiting the freedom of the individual research worker.

A third assumption is that where prioritization is undertaken, then it must be possible to reach the stage where resource can be moved from one part of the program to another. To this end, the Council has established its Advisory Panels and has constructed the policy making cycle described in this report. Although it would be premature to judge its effectiveness after a single cycle there is no doubt that important freedoms have been won by moving away from the lobby mode of the former policy making systems. In terms of policy formation and project selection, clear benefits can already be identified, including higher success rates, faster response times, more open and participative peer review, open and succinct priority statements, absence of submission deadlines and a sympathy of the overall system to interdisciplinary themes with the removal of Committee territories. At the same time, there are continuing difficulties; these include misgivings about the inherent conservation of peer review and concerns about the role of Program Managers in the preparation of advice for the two Council Panels.

Interdisciplinarity in America 1949–1999 Experiences of a Proactive Champion of the Cause ▶▶

Rustum Roy
Materials Research Laboratory
The Pennsylvania State University

"The momentum which impels investigation to dissociate indefinitely into particular problems, the pulverization of research, makes necessary a compensative control-as in any healthy organization-which is to be furnished by a force pulling in the opposite direction, constraining centrifugal science into a wholesome organization...the selection of professors will depend not on their rank as investigators but on their talent for synthesis."

—*Jose Ortega y Gasset,*
"Mission of the University"

Introduction and Background
The title of this paper is clearly too expansive without the subtitle. I start by noting that early in my career I intuitively became committed–without much analysis–to a wholist, integrative approach to scholarly work and to life. The logic of Ortega's thesis quoted above is so absolute, that the reductionist fragmentation of knowledge absolutely demands in a mathematical, physical sense, a "compensative control" by interdisciplinary, interactive, integrative forces. As will appear in the record below I have pursued over my 50 year long academic career a series of actions on a wide

range of platforms to achieve some of this "synthesis". This paper attempts to draw some lessons for the future from this experience.

This author fits Shakespeare's third category in that: he had "interdisciplinarity thrust on" him by fate. I was trained as a chemist B.Sc (1942) and M.Sc in (Physical) Chemistry (1944) in India. But in the U.S., between 1946 and 1949, while working more or less in the same area my major was changed from Chemistry to Ceramics; although I was able to take the same courses to get a Ph.D. degree in either. I was then appointed as a Research Associate in Mineralogy, and a year later in 1950, I was appointed as the first Assistant Professor of Geochemistry–a neodiscipline–anywhere. The fungibility of the category "discipline," and the arbitrariness of what is included within a department or discipline was thus impressed on me early.

Lesson: Disciplines and departments are never clearly defined.

In fact I was a synthetic inorganic chemist, working on phase equilibria and establishing an expertise in crystal chemistry, which made my research very relevant to physicists and ceramic scientists and engineers. In the 1949-59 decade I had very close interaction with the physics community (on campus under R. Pepinsky, and in industries such as Bell Labs) both for ferroelectric properties and the many dozens of new phases we were making. Our discoveries of the sol-gel and modification of very high pressure, processes brought us many interactive collaborators from all over the world. Under the departmental label "Geochemistry" we have several senior physicists and were training dozens of Ph.D's who were in great demand in the major "materials" departments or industries (Bell Labs, GE, etc.).

Lesson: Really Frontier science will always need and find collaboration.

The Pattern of Response to the Need for I³R

Starting from this background over the next two or three decades there emerged a series of innovations in education and research, which demanded some new ways for new groupings of scientists to interact and work together. In retrospect only perhaps one might discern a 'gestalt' in the series of events—conferences, workshops, journals, etc.—in my existential "research" on interdisciplinary organization, which I list in Table I.

Table 1.

PSU-MRL ROLE IN CATALYZING INTERDISCIPLINARITY	
1953-59	Local Experience/Experiments
1959	Established First Interdisciplinary Degree in Materials • It Is Still Run by a Univ. Wide Committee
1962	• Established MRL and Umbrella Org. (ISE) • Created Industrial Coupling Program in MRL
1965	Materials Advisory Panel of Gov.'s Science Advisory Committee (6 Corporate V.P.'s and 6 University Leaders from State) Meeting Quarterly to Advise Gov. of PA
1966	Evaluation of ARPA IDMRLS (for ARPA)
1966	1ˢᵗ Intnl. "Materials" Meeting (on "Characterization of Materials") • Triggered U.S. National Academy to Deal with Topic • Triggered MRS Formation
1968	2ⁿᵈ Intnl. "Characterization of Materials" Meeting in Rochester
1969	1ˢᵗ Intnl. *Policy* Meeting on MSE and How It Fares. • COSMAT Triggered. (book provided)
1971	1ˢᵗ National Meeting: University-Industry Interaction. Ed David, Sci Advisor to Pres. Nixon, keynoter.
1973	COSMAT Report; W.O. Baker, M. Cohen, Chairs; (Roy Chairs Univ. Education Committee)
1973	First Meeting of MRS (at Penn State)
1974-Present	Continuing Local Experimentation on Univ. – Industry Interaction – with ≈ 100 Companies Annually – Individually; Consortia; NSF Centers etc.
1960's	HQ for Science and Art Interaction on Campus and Nationally
1970's	Host for STS Program on Campus and Nationally

New Interdisciplinary Graduate Degree Established

In response to the obviously anomalous situation and the simple necessity for truth in labeling in our degrees programs, in 1959-60 we established, at the Pennsylvania State University, under my chairmanship an interdisciplinary graduate major in materials called, "Solid State Technology."

The key characteristics of this first *educational* interdisciplinary program have proven to be very significant, and are as relevant today as forty years ago. First, the degree program reported directly to the Dean of the Graduate school; it was not in any department or college. The supervising faculty were drawn from seven (or more) departments, Geochemistry, Physics, Chemistry, Ceramics, Metallurgy, Engineering Science, Electrical Engineering. This group of faculty established the core requirements, which every student had to take, and administered the usual candidacy comprehensive and final examinations. The enormous advantage of this program was that we could accept graduate students from a wide range of undergraduate majors interested in research on materials. These students would not have to take the 15-30 make-up credits that the typical department requires if one comes from another major. The intellectual invention was: carefully defining the core–at once broad enough to justify the title the University confers, and to allow the students to fill it out with other course work, and deep enough in some aspect to be able to enter the research world at an advantage. I taught classes in Crystal Chemistry to 50-60 graduate students from major U.S. Universities, with backgrounds in a dozen fields (the Physics and EE majors had long since forgotten the periodic table!). The Ph.D. degree in Solid State Technology was an immediate success. By 1968, less than a decade after it started, when I stepped down as chair, we graduated 22 Ph.D.'s *per year* (making "SST" among the two or three largest Ph.D. producing units in the University). This genuinely interdepartmental and interdisciplinary degree program (today called "Materials") has continued at a level of about 100 students enrolled. It has served as a model for several later degree programs from the Ivy Leagues to Texas Universities. In parallel with this wider use of the term interdisciplinary i.e. involving most departments in science and engineering colleges has emerged the much more common "combination" of ceramics, metals and polymer units into a new department or discipline often called "Materials Science and Engineering (MSE)." Several dozen such limited-interdisciplinary academic programs formed over the last 30 years. The intellectual success of the concept is not assured, as polymers are rarely fully integrated into so-called *MSE departments.* (See next section) Some MSE programs are being integrated into other groupings such as mechanical engineering. In retrospect, the two key elements in a genuinely interdisciplinary degree are: (a) to do the hard work to define the core course requirements; (b) to have the interdisciplinary group of faculty report to the Graduate Dean directly, not through any department.

Early in the conceptualization of the field I had concluded that one must keep nomenclature very clear. This was spelled out most fully at the 1972 conference "Materials Science and Engineering MSE in the U.S." (see above) where the absolutely essential distinction between MSE (as the new re-integrated *teaching discipline*) and Materials Research (as the *interdisciplinary activity* in which engineers/scientists from many disciplines could participate) was shown. Hence, we have consistently used Materials RESEARCH as the prefix for interdisciplinary labs, journals, and society. Regrettably, sloppy and indiscriminate use of "MSE" for interdisciplinary research (even by the U.S. National Academy in its major reports) leaves no room for the 100 University departments or disciplines which also use MSE to describe an obviously narrower field of activity.

Lesson: Accurate use of terminology is as important as getting our equations correct.

Journals
In the early sixties it had become clear that one way to focus the attention of any subset of the science community was to create a professional scientific journal and a professional society. In 1965 my colleague Prof. H. K. Hemisch and I approached Robert Maxwell, founder of Pergamon Press, to start such a new journal in the budding interdisciplinary field of materials

research. Thus was born the first such journal "Materials Research Bulletin." In fact, within the year 1966, two other new materials journals also appeared. Since then literally dozens of journals in the general area of materials research, and one in Materials Education have been started and done well. Unfortunately many of these continued or enhanced the fragmentation instead of unification of the field.

Lesson: Journals are not of any value for this cause.

Creating the World's First Independent Interdisciplinary Materials Research Laboratory

The second organizational innovation (see Table I) towards interdisciplinarity took placed in 1962. By then about a dozen Universities had obtained ARPA IDMRL *contracts*, and created various Administrative structures to run the contractual programs—under names such as Center for Materials Research, Materials Research Center, Materials Research Laboratory, etc. etc., Penn State had not succeeded in these first rounds of proposals, but in the process of preparing the proposals, the arguments were found to be so compelling that the Vice President for Research, E. F. Osborn, himself a distinguished materials researcher, and President Eric Walker (soon to be the first President of the nascent National Academy of Engineering) made the decision that if interdisciplinarity is essential for doing good materials research, our University should not depend on an on and off Federal Agency to decide on our structures. Hence, they established at Penn State the first Materials Research Laboratory in the country, and world, in 1962, with no connection to any government agency. (Of course, the faculty would seek funds for research, but *from* the MRL, *not to create* a MRL.) I was appointed the first Director. Moreover, (uniquely in the world, even nearly 40 years later) the University made the MRL a tenure-granting unit, and a unit where a faculty member could hold her/his primary affiliation. E. F. Osborn, having thus committed the University to the interdisciplinary path in research, then created an intermediate level of research management structure: (then named) the "Institute for Science and Engineering" to house (like a college) a set of similar interdisciplinary programs (e.g. in the Environment, Transportation, Art and Humanities Studies, etc.). This structure continues more or less as such to this day. However, a major change occurred about 1971 with the change of President and Vice President. The MRL lost its right to grant tenure. It became clear to me, by experience, that without a champion at the very top, the disciplines, departments and colleges naturally chipped away at the interdisciplinary units. Thus the third important sine qua non for an effectively functioning "interdisciplinary" unit, is active, public, support from the V.P. for Research and the President.

Lesson: The hallmark of success in institutionalizing interactive research is for the University to create its own structures and units irrespective of any government RFP's or contracts. A champion of the cause above the Deans' level is essential.

Use of Scientific Meetings on Interdisciplinary Topics
By the mid sixties it had become clear to advocates of more interactive research across the different classes of materials–metals, ceramics, polymers, etc.–that the professional societies based on those materials classes would not hold meetings on "interdisciplinary" topics.

We therefore launched a campaign to organize a series of such meetings to bring together the budding interactive research community. The first such technical meeting was the First International Meeting on "Characterization of Materials" held at Penn State in 1966. It was at this meeting that informal discussions were started on the possibility of starting a new professional society. This was followed by a second meeting in 1968 at Rochester, NY, at which I took an official ballot on how many wanted a new society. The result was 30% for, 70% against!! In 1967 I had already approached Dr. Frederick Seitz, (himself a distinguished "materials scientist") who was President of the U.S. National Academy of Sciences. He appointed an adhoc committee, to see if the Academy could help steer the field. Some years later the committee presented these options: 1) Form a Federation of Materials Societies, 2) Form a new society, 3) Start a national "study" of the field. As events turned out later, all three were done.

In 1969 I organized and chaired the first International meeting to evaluate the status of the field of "Materials." It was held at Penn State's M.R.L. and the attendees were a Who's Who of the world's leading materials scientists. Thoughtful papers describing not only the status of separate subfields all over the world, but the differences between interdisciplinary teaching and research, etc., appeared under the title "Materials Science and Engineering in the U.S."

In 1972 we also organized the very first Conference on University-Industry Cooperative Research. High-level agency representation, including Edward David, Science Advisor to President Nixon, attended. Both these last two meetings were sponsored by the State's Materials Advisory Panel of the Governor's Science Advisory Committee, bypassing the National Agencies and Academy and their lengthy bureaucratic procedures. The first *international* collaboration event with Japan took place in Tokyo, in 1969, when I co-chaired the U.S. delegation for the Japanese-American "International Cooperation in Ceramic", Seminar.

Lesson: Meetings are a necessary but not sufficient vehicle to create interdisciplinary interaction.

New Societies

The arguments for starting a new society–eventually, accurately named the Materials Research Society–were mounting from 1967 onwards, as noted above. They included:

1. It would provide a regular meeting venue for the large number of scientists actually participating in interdisciplinary activities in industry

and government labs, and allegedly in most Universities, to present their results.

2. It would help create a new focused community of those committed to interdisciplinarity.

3. Such a community would honor and recognize persons and research where interdisciplinarity was a key parameter.

The Materials Research Society was formally launched in May 1973 at its inaugural meeting at Penn State. Some 300 persons attended. The subject again (as in the 1966 and 1968 Characterization Conferences) was a topical one, which cut across all classes of materials: Phase Transitions.

The handful of founders made certain key "inventions" which are now universally copied. First, the sessions were *always* structured around research themes or new areas, not following the disciplinary substructures at all. Second, we proactively sought out new areas, obviously those sparked by new discoveries. Third, I insisted that MRS always balance out the separate 'symposia' by sessions cutting across them, where senior figures would review major fields, or deal with wider issues such as education or policy. Another key was to have societal awards and recognition specifically for interdisciplinary achievements. Regrettably, with time, this key feature seems to have been muted within MRS itself. For the first ten years of its life the headquarters of the MRS was in our Penn State Materials Research Laboratory where I was Director. The society was no runaway success in its early years. Most scientists from Universities, albeit 20-25 had some kind of official interdisciplinary materials organization, paid very little attention to it, preferring to attend their traditional disciplinary societies. Fortunately, industry and government researchers thought otherwise. Indeed, MRS was greatly helped by the coincidence that by 1978, two major new fields of research, were being pursued very actively by agencies: radioactive waste management and laser processing of materials. Symposia on each of these topics brought out a few hundred to each and by 1983, with one half time staff member, Mr. E. Hawk, our tiny MRS-HQ was running, with the vigorous help of symposium and meeting chairs, national meetings of 1500 attendees. The time had come for MRS to move from nucleation to growth, to set up its own infrastructure. This transition was effected very smoothly and the society has gone on to a dominant position in the modern materials field. The success in the growth phase owes a great deal especially to a handful of government national laboratory scientists and several senior industrial research leaders who were deeply committed to the cause. It is significant that for nearly the first twenty years University personnel (excepting Prof. Gatos at MIT and the author) played minor roles in the society. Today the Materials Research Society is recognized world wide as the premier centripetal force in the field. Its meetings are the most significant gatherings of the "materials-research" community.

Lesson: Because of the funding structure and social habits, a high visibility scientific meeting run regularly by a new society in a new field can be a very powerful glue for a new field.

Inter-institutional Interaction: Activities at the State Level

In 1965 the Commonwealth of Pennsylvania, under Governor W.W. Scranton, created this country's first State-level Governor's Science Advisory Committee (GSAC). In addition, the visionary Secretary of Commerce, Mr. Clifford Jones, foresaw the potential of getting the Universities and Industries to work together to seize technological opportunities. Pennsylvania, the heartland of the nation's steel, aluminum and ceramics industries obviously had a special interest in the *materials* industries. Hence, he created a Materials Advisory Panel of the GSAC; and as one of the latter's members, I was appointed Chairman.

This committee of six Pennsylvania University Materials Research Laboratories (or departments) and ten Vice Presidents of major Materials companies (the backbone of Pennsylvanian industry) met several times a year, analyzed situations and wrote reports to the Governor, on encouraging university-industry collaboration, on the most promising materials technologies for the economy of PA (Wood was ranked far above semi-conductors). This proved in its 15-year life to be an effective catalyst for enhancing inter-sectoral and inter-institutional interaction and research. In sharp distinction from the national level, National Research Council, and its Materials Advisory Board, 100% of the GSAC-MAP work was started and directed by the committee itself, not at some agency's request.

I³R: University-Industry Interaction

Experience is the greatest teacher. My professor and mentor, E. F. Osborn, came to Penn State from the Eastman Kodak Co. where he had worked on new optical glasses. With a tiny grant and the loan of some large platinum crucibles he continued working on the fundamental phase equilibria of beryllium fluoride based glasses. With another tiny grant from the Harbison Walker Co., he started work on hydrothermal growth of refractories, and soon he got a modest grant from the Bethlehem Steel Company to study the fundamental phase equilibria relevant to steel plant refractories. That, of course, meant all the oxide systems fundamental to much of modern ceramics of the time. That industry grant continued for 27 years without interruption: it made our lab the University world's HQ for the fundamental science of determining phase equilibria. But, this basic work was not esoteric; it was actually translated into modification of the composition of slags in the blast furnace. Industry can generate very basic science, and yes, you can put it to use for that industry.

The years were 1946-48. There was only one Federal agency and no Federal "gravy train." In 1948 Osborn obtained the very first Federal grant to Penn State from the ONR and I was the first person hired on any Federal research contract in the University. Hence the coupling of Universities to Industry was much more intense and real in that period. It was obviously not a *new* idea to those who entered the professional world before say, 1955. And the dominance of government funding itself was responsible for breaking the natural interaction.

When I became Director of the local MRL in 1962, I immediately started for the whole lab what we called the "Industrial Coupling Program" (ICP). I argued that *the most important gain to us in the University was industry's distilled experience on identifying the most significant problems.* They led us to the ore, where we could do our most *influential* basic research. Yet on campus this move of starting an ICP was opposed or belittled, saying that we would be doing "testing" for industry! Thirty years later the MRL not only had by far the highest percentage of support from industry of any MRL on the campus or in the country, it held Penn State's two successive top revenue-earning patents. Our MRL has averaged support from *over* 100 companies/year for decades. We have half a dozen consortia, individual grants, and individual contracts, with reasonable protection of proprietary rights for the sponsors, etc.

Obviously we have made University-Industry interaction work. Our goal in such work has always been to get some support for the more long term end of research, especially on societally significant problems, *never* to make money form patents and licenses, etc. The latter as it happened has been a by-product. Our really new discoveries are eagerly and objectively examined by our industrial colleagues, who visit our laboratories, examine our results, take away samples. That is true peer review. Amazingly, time after time, even when *several* companies have funded our work, the government agencies peer-review opines "It cannot work."

By far the biggest hurdle to interaction was introduced after the mid 1980's with the University's sudden interest in exploiting its "Intellectual Property" and putting up all kinds of difficulties in working with industry. Because of our successful track record we have been able to work the system as not to stymie our work altogether, but an awful lot of time and effort have been wasted. Details on this University's handling of the problem may be obtained by written enquiry from the author. It cannot be presented easily because it changes so frequently.

Lesson: What have we learned about effective coupling between industry and universities? It is almost 100% a matter of attitude. We believe (and act on that belief), that industrial partners are our equals in every way. Their task is much more difficult than ours. All we need to do is to do something new. They must do that and make it pay. Moreover, they can often guide us to very significant problems; or industry can objectively examine our discoveries, for which we can jointly, or alone, get government funding.

Interdisciplinarity Outside Materials Research

The Materials Research Laboratory because of its commitment to interactive research and teaching became a kind of champion of, or prime mover in, several other interdisciplinary activities on this campus and in the nation. I will mention very briefly three very different areas: Science and Art; Science, Technology and Society; and very recently, Integrative Medicine.

Science, Art and Humanities

From the earliest years of the laboratory, the faculty's own interests extended to a proactive reaching out to establish courses, seminars, exhibits, national and local competitions, and finally a museum within the building, of Science and Art. This was well received by the University's Arts community and some visitors but had little impact on most scientists, albeit they recognized in the MRL a radically distinct esthetic milieu compared to other labs.

Our concern for "interaction" in a Land Grant Institution led us to focus on the "public" as a major constituency. The Public Understanding of Technology was a large NSF funded project in 1970 in the MRL on a topic, which has in the last two years suddenly been rediscovered at the highest reaches of the engineering community. A wide variety of activities included workshops at church and union meetings, producing 3 PBS-TV mini-series etc.

Science, Technology and Society (STS)

While Materials Research had a bandwidth, which included, say, half of science and engineering, it was soon eclipsed with regard to breadth by a much more ambitious effort at the integration of knowledge across Science, Technology and Society. In 1969-70 I appealed to the local University Senate to create such a program in Science, Technology and Society to help our undergraduates understand their world which is now dominated by Science and Technology. Penn State, along with Cornell, Stanford, and SUNY Stonybrook, pioneered in establishing the program, degrees, research, etc. in STS. I was an early Chair and first Director of Penn State's STS Program. In funding, STS was the very opposite of Materials–there is virtually none. Yet, amazingly, today on this nation's major campuses there are about as many interdisciplinary STS programs as there are IDMRL's–several dozen. Moreover, STS has penetrated the K-12 field to an unexpected extent, many States requiring it and thousands of schools offering it. Prof. R. Yager, current President of the National Association, will provide more data on this in his paper.

STS followed the strategies learned from the materials research field:

- Hold annual national meetings.
- Start a national (and international) society.
- Start new journals and provide teaching modules.
- Work vigorously with political and social leaders to gain acceptance.

The results were *similar* to those in Materials except that STS is largely a teaching effort, with only modest research activities, and hence the speed of "STS Community" build up was an order of magnitude slower, since Federal funds subsidizing national meetings were minuscule.

As a teaching element, interdisciplinary STS encounters the same great resistance from the existing fields. For example, since in K-12, "science" is equated with physics, chemistry and biology, or social science, how can one "insert" into the curriculum a new integrative science usurper? In college the situation is *somewhat* better because of the demands that engineers have for more humanities, etc.

But overall, the most obvious and compelling arguments are never challenged, but the change toward the new integrationist learning is glacially slow.

Integrative Medicine

This is the most ambitious and radical attempt to integrate within medical education the insights, methodologies, and results of many alternative modalities. There is, however, a major favorable element in the case: the demand by the public and young physicians is so large, that it will drive the system. I am now formally affiliated with this effort through my appointment at the University of Arizona.

Although I had been reading the works of Illich, Cousins, Rodale, etc. in our STS work for years and have been studying "alternative therapies" for some years, it is only very recently that I have become formally involved in the field. Dr. Tracy Gaudet, Director of the University of Arizona's Program in Integrative Medicine gives the details of their work in her paper.

Summary of Learnings from Experience, re I³R:

First, those who practice I³R must be clear of the intellectual and moral case for their own position, and be able to articulate it. I³R should never be presented as a strategic choice to get more funding, although much of the flimsy "interdisciplinary" infrastructure that exists is built on just such motivations. And, indeed, the *motivation* of the Agencies to encourage and facilitate I³R is obviously genuine and essential. However based on the track record uncovered in this conference, the strategies for achieving the interactive research goals clearly need some work.

Second, it is in fact a return to core values of science, of society, for some of the Land Grant Universities. We have lived through a few decades where the value of isolated, detached, disciplinary, often esoteric, theoretical only, research was grossly exaggerated. It is a strange time when "Edisonian" is used as an epithet by scientists reviewing research in America (!), whereas simulations, "first principle" calculations however disconnected from reality and experiment are prized. It surely behooves us to ask again: What are first principles? Aristotle in his Nicomachean Ethics (I, vii, 17-22) may help. He has the following to say:

*"Nor again must we in all matters alike demand an explanation of the reason why things are what they are; in some cases it is enough if the fact that they are so is satisfactorily established. This is the case with first principles; **and the fact is the primary thing–it is a first principle**. And principles are studied–some by induction, others by perception, others by some form of habituation, and also others otherwise; so we must endeavor to arrive at the principles of each kind in their natural manner, and must also be careful to define them correctly."*

And later criticizing a colleague's work on the reproduction of bees Aristotle includes an important caveat:

*"The **facts** have not yet been sufficiently ascertained. And if at any future time they are ascertained, **then credence must be given to the direct evidence** rather than to the theories; and to the theories also, provided that the results, which they show, agree with what is observed!!"*

These wise words are as relevant to a whole variety of I³R research. The decision by U.S. industry to abandon speculative research not connected to their "facts" (products) is a most significant move, in this connection. They apply to the most startling and interesting discoveries in materials research in our previous experience: from the "theoretically impossible" high Tc superconductors, to totally new materials processing techniques such as directed oxidation, multiple pulsed lasers, microwave synthesis, as well as the entire field of integrative medicine. For the last several decades the observations and facts (e.g. of healing by unconventional paths) that Aristotle argued for, have been totally ignored because "credence" was given to the medical establishments theories rather than "the direct evidence." An interdisciplinary approach avoids the limitations that a narrow discipline imposes.

The New Epistemology: Discrediting the Linear Theory
The ludicrous proposition, the so-called linear theory, that *basic science leads to applied science leads to engineering and technology*, regrettably absorbed as gospel by probably the large majority of scientists and engineers trained after, say, 1960 has at last been thoroughly discredited among all students and practitioners of science policy. This theory is, of course, the necessary result of the reductionist absurdity that the whole can be made up from parts. It is first essential therefore, that this entire set of ideas and values be openly discussed and discredited, because it is impossible to substitute the new worldview until this is done. The new epistemology for pull-science, connects science to technology and human needs. It starts with felt needs, moves to careful observations, especially of "anomalies" and exceptions and curiosities, of the whole, suggest new opportunities. From such "observations" one works back to technology, which pulls out the relevant applied science. This is followed by technology, which pulls out the relevant applied science. This is followed by careful analysis and optimization and perhaps in a few cases pull out some science, new understanding and maybe "new principles." The two epistemologies are compared in Fig. 1. The classical view was presented in the National Academies, 1974 COSMAT report on the Status of the materials world; the new version is the key to STS pedagogy.

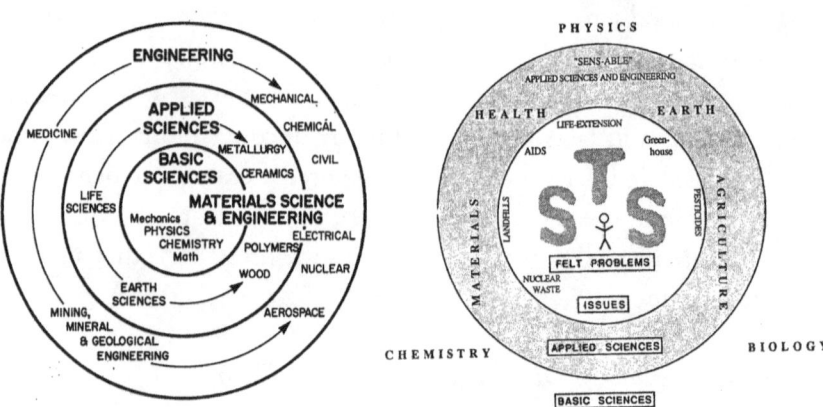

Figure 1: Two epistemologies compared. On the left is the now discredited linear *theory* of knowledge-flow, which is still firmly the belief of most scientists. On the right is the new need driven theory for science/technology.

There never were any important examples to support the "linear theory." At the best, it was a rare exception. Aristotle's wisdom now over 2300 years old must be taken seriously by the science community. **"Credence must be given to the direct evidence."** From the steam engine to the transistor or high Tc superconductor, the direct evidence says, "Science has followed from important inventions." As the "time to market" drives inventions, science must learn to be a "fast follower" of the unexpected discovery or invention not a doubting Thomas isolated on the sidelines. (See Fig. 2.) The major recent basic science discoveries in my own research group (microwave processing and pulsed laser processing) have all come by *starting* with work very close to the final product, usually carried out with an industrial partner. A step-function improvement there *demands* (and supporters are willing to pay for) an understanding to guide further improvements. This is obviously another major justification for intimate Industry-University coupling.

Failed Linear Theory	*University* Idea→Theory→Experiment→(Mat?)		*Industry* Development→Product
Third Millennium Reality	*Industry* *(to others)* Real Material/ Process Discovered Here ◀	*University* Validate, Broaden, Develop Understanding via Broad Gauge Scouting Worldwide ▶ ◀	*Industry* Goal: Better or ▶More New Materials

Figure 2. Universities do backwards Integration from discoveries: This is a wholly new role: i.e. do the science behind discoveries made elsewhere.

The Essential Re-structuring of the University to favor I³R

In 1977, in a cover story article in Chemical/Engineering News (The reader is referred for the full argument and details to the appendix which reproduces this article). I proposed that at least some Universities should be structured along the complementary integrationist system instead of the reductionist discipline model.

At present out of permanent vertical units called disciplines or departments one aggregates horizontal temporary dependent variable interdisciplinary units created and maintained as semi-permeable membranes. This can be seen as the exact obverse of the profoundly different arrangement, which we deal with next.

		Chem	Biology	Electrical Engineering	Engineering Science	Physics	Etc.
Temporary (5 yr) aggregation of research teams in various topics	Food Technology	✔	✔ ✔		✔ ✔		✔
	Information Technology			✔	✔		✔ ✔
	New Electronics	✔		✔		✔	✔

Figure 3. Present Dominant Substructure of Universities

Interdisciplinarity in America Universities Beyond 2000

That sounds like an oxymoron, but it need not be. In all non-academic scientific organizations there is already a strong element of matrix management. The only question is which are the dependent variables—the temporary structures? Why? The present approach in 100% of U.S. University can be represented above in Fig. 3.

In the future the societally-needed goals, objectives or tasks will form the "semi-permanent" vertical structures in a university, and the "Teaching units (=disciplines) the new dependent variables will be assembled by buying time from persons in those units trained and capable of teaching in a particular field. Further, both vertical and horizontal separators, will be more transparent to each other and towards industry and public sectors. (See Fig. 4.)

Societal Need Aggregations

These would be major permanent subdivisions

Departments (These would be flexible units for teaching)		Food	Health	Earth Resources	Information Technology	Etc.
	Chemistry	○	○	○		●
	Physics			○	○	●
	Electrical Engr					
	Materials Engr	○	○	○	○	
	Biology	○	○	○		●
	Mechanical Engr	○		○	○	●

Figure 4. Proposed Structure Based on Major Needs of Society

Reforming Present University Structures to Support *βR*
Mandated Joint Appointments

A relatively simple and achievable goal for almost any university would be to require that within say a 10 year period, 25% of every department's faculty would have to hold permanent joint appointments in other departments or research units.

Tenure in Both Departments and Interdisciplinarity Units

In selected research or societally-relevant fields (different Universities would choose different fields) a number of tenured positions would be assigned to the most permanent and significant I³R units. Obviously, only shibboleths can be used to claim that about 100 sacrosanct concatenations labeled departments qualify as tenure homes, while say 10 others, which are interdisciplinary do not!! In my University, as in many, there are departments of Geology, Chemistry and Geochemistry. In 1950 the last named didn't exist. If it were called an Interdisciplinary unit (either teaching a research) it would not have been able to be a tenuring unit. This is a *very* simple move, which any administration could implement, thereby sending a strong pro I³R signal to the community.

Using I³R Activities as Necessary Qualifications for "Super-Grades of Professors"

Policies could be instituted to encourage interactive teaching and research by *requiring some track record specifically in some interdisciplinary areas* to qualify for named chairs, for distinguished Professor rank, and finally for Emeritus rank. This would provide a normal continuing incentive towards I³R, easily compatible with present structures.

Leadership

There can be absolutely no substitute for continuing leadership supporting I³R by the President, Provost and V.P. Research. The disciplines have an entire supportive infrastructure of appointments, promotions, salaries, up through the Deans, with both powers of appointment and promotion. Without positive interventions by such leaders on behalf of I³R work, interdisciplinarity cannot make it in the sea of academic's disciplinarity.

The Interdisciplinary Imperative

Chancellor Bugliarello raises in his very carefully architected paper all the key arguments for interdisciplinarity. The "absolute" argument (also alluded to by him) however is the one I referred to in my 1977 paper. It has to do with the finite capacity of the human brain and person. Figure 5 shows how this fixed, finite, capacity for negentropic work, or organization of information into knowledge, superimposed on the amount of 'knowledge' available in say 1750, 1900, 1950 and 2000, in various 'disciplines.' The choice facing educators in every scientific "discipline" is only: *What* content should we as a faculty choose to define as that which every person educated to a certain level in our field needs to know. Few today could support the idea of choosing a small subset of knowledge all within one discipline. The choice, to overlap 2 or 3 fields, looks very obvious—hence interdisciplinarity will inevitably become the normal "discipline" early in the new millennium.

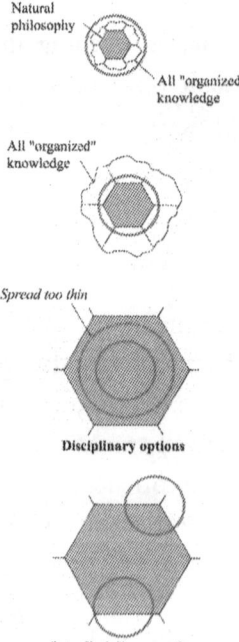

The rationale for interdisciplinary units on campus

Natural philosophy

All "organized" knowledge

All "organized" knowledge

Spread too thin

Disciplinary options

Interdisciplinary options

Figure 5. The changing ratio of fixed human negentropic capacity, to expanding knowledge in all fields The circle represents the fixed brain capacity, while the area of the hexagons represent the knowledge in a particular field or discipline. Once the organized knowledge greatly exceeds the human capacity, the only choice is how to place the circle over contiguous fields of knowledge. In most cases the choice will cut across two or three contiguous fields, the interdisciplinary option. But some may choose to stay within one 'discipline', and hence become subdisciplinary, or as the larger circle shows sacrifice depth for some increased breadth.

Acknowledgements
Materials Research and all I³ Research were extremely fortunate to have had a risk-taking Vice President and President at Penn State in the formative years of the sixties. No less fortunate were we in the support risk-taking administrators in a few agencies—especially the Office of Naval Research—and a few private Foundations provided over there four decades. I also wish to record my gratitude my colleagues Professors L.E. Cross, R.E. Newnham, D. M. Roy, G.R. Barsch, and K. Vedam for helping work out the realities of our own local experience with an interdisciplinary culture in a disciplinary setting, as a small group of friends.

Two Examples of How Interdisciplinary Academic Programs Evolve ▶▶

Mohammad A. Karim
Head, Department of Electrical Engineering,
The University of Tennessee, Knoxville

Katy E. Marre
Associate Vice President for Graduate Studies and Research
University of Dayton

Joyce Jentoft
Associate Provost and Dean of Graduate Studies
Case Western Reserve University

This paper is a reflection on observed changes in two disciplines in which interdisciplinary work has evolved into new disciplines and resulted in new specialized disciplinary choices for students. We consider two case studies: Electro-Optics and Computer Engineering with a view to propose changes in the way we train students engaged in interdisciplinary projects and how we develop, support, and evaluate their contributions in the programs.

While the image of the lone researcher is still the most common format of the knowledge-driver, the various research-sponsoring agencies, for a while, have been rewarding those research efforts that involve multiple investigators. The most preferable amongst them are those that involve investigators of multiple disciplinary backgrounds often originating from multiple units, organizations, and institutions. This trend is in evidence also in the ways small high-tech companies are being created and sustained as they contribute to the nation's ability to develop, adopt, and diffuse new technologies. National Science Foundation (NSF) has identified automation, biotechnology, hardware, advanced materials, photonics and optics, software and telecommunications as the seven critical areas [1]. Table 1 lists job growth in the seven most critical areas in US. Every one of these seven areas is multidisciplinary in nature. In the purely academic world, however, such multidisciplinary and/or interdisciplinary research initiatives do not yet appear to influence academic programs in a major way. The typical undergraduate or graduate student still seems to get funneled through strictly designed academic departments.

Table 1. US job growth in the seven most critical areas [1]

Field	Job growth past year	Jobs added past year	Sales/employee	% exporting over 10% of sales
Automation	12.6	58,471	$126,330	40.7
Biotechnology	10.6	16,468	$86,063	43.9
Hardware	14.5	148,304	$172,752	42.9
Software	13.9	103,479	$107,992	32.1
Advanced Materials	5.8	32,452	$155,198	44.4
Photonics & optics	8.4	27,654	$103,592	36.8
Telecommunications	12.7	61,280	$117,679	43.9

A significant number of studies have identified the scope of the challenges for the academic community. These challenges, according to the American Society of Engineering Education (ASEE), include the need to attract a greater diversity of students, as well as the need to shift from a technology policy strongly focused on national security to one aimed more diffusely at international economic competitiveness, communications, and sustainable development [2].

In this paper, we analyze the rationale, implementation process, and outcomes of initiating two interdisciplinary academic programs involving one or more academic disciplines that reside within the traditional boundaries of engineering and sciences. Both of these interdisciplinary academic programs respond directly to the critical areas as identified by the NSF [1]. Each of these interdisciplinary academic program initiatives is analyzed from the perspective of the experiences of the U.S. universities in general, and from the perspective of the University of Dayton, in particular. As a general observation, it is puzzlement that while "interdisciplinary" suggests a synthesis through interactions between and among disciplines, it actually provides *more* specializations to graduate students. In the two specific instances we discuss, there have been evolutions in the disciplines which have resulted in new academic specializations, serving new and different groups of students at both the graduate and undergraduate levels pursuing newly designed degree programs or interdisciplinary research projects.

In Electro-Optics

Electro-Optics is now becoming a discipline by its own right. It has evolved from being a sub-discipline of physics to becoming an applied discipline by engaging the powerful tools and techniques cultivated in engineering, in general, and electrical engineering, in particular. As is obvious, this evolution is in full agreement with the market force data such as that already identified in Table 1. For historic and philosophical reasons, however, physics and electrical engineering programs in U.S. institutions of higher learning have remained under two different administrative structures: science and engineering. Not surprisingly, therefore, few interdisciplinary programs of any kind come easily to fruition between programs of differing administrative structures. The usual reasons for which such programs are not cultivated are: existing competition across institutional boundaries; infighting from the established disciplines from within the structure itself; fear that such interdisciplinary programs may drain away resources from well-established disciplines; and lack of resolve on part of the optics community to formalize optics education outside the traditional realms of physics and electrical engineering. But the job market, which is often ahead of institutions of higher learning in planning, indicates many interdisciplinary trends. For example, more than 50% of the jobs suitable for a Ph.D. in optical science and/or engineering remain unoccupied in the U.S. since 1988 [3]. At the master's level, the situation is somewhat better, but for 20% of such jobs there are no takers. The overwhelming majority of those who end up taking these jobs are typically graduates of either physics or electrical engineering programs. Unfortunately, that is not the most preferred solution but, for all practical purposes, it is the only stopgap solution. Employees do recognize that a graduate of either physics or electrical engineering programs can function well with certain aspects of optics, but not with the whole. The need for cultivation of this interdisciplinary discipline is obvious.

In 1990 alone, over 9700 undergraduates were enrolled in optics courses [4]. These courses are typically available as electives for the physics or electrical engineering seniors. The overwhelming majority of these students who chose further study, however, did not major in optics. The professional community, made up mostly of physicists and electrical engineers who are exploring different aspects of optics, is still hesitant about the very name of this interdisciplinary discipline. Some of the more common names are "optical sciences," "optical engineering," "Electro-optics," and "applied optics." Consequently, even in those U.S. institutions where it has become a new discipline, some have chosen it to be a part of sciences while the others have configured it as a part of engineering. In an era of economic belt-tightening, however, institutions are faced with this problem: the faculty that can serve as the seed of such a newer venture are currently housed in two or more traditional departments and colleges; and optics education is still an expensive and mostly graduate venture.

We can categorize institutions having academic programs/departments with teaching and research emphasis in optics into three types. *Type A* institutions are those which offer stand-alone degrees in optics. *Type B* institutions are those that their traditional physics and/or electrical engineering departments offer a specialization or concentration in optics. *Type C* institutions are those wherein physics and/or electrical engineering departments offer only one or more courses in optics but is unable to offer much of a concentration in optics.

Of the 237 non-trivial U.S. programs listed in *The Guide to Optics Courses and Programs in North American Colleges and Universities*, Type A, Type B, and Type C programs respectively number 8, 75, and 154 [4,5]. Table 2 lists the Type A institutions and they're various characteristics. In three cases (Arizona, Dayton, and Rochester), the new interdisciplinary academic unit is administered out of a budgetary unit completely separate from both physics and electrical engineering (EE). In the remaining cases, the new degree program is an extension of the other existing traditional program(s) in either physics or electrical engineering. It is also interesting to note that the majority of these institutions are co-located regionally with major optics industries (as in the cases of Arizona, Central Florida, and Rochester) or government defense laboratories (as in the cases of Alabama in Huntsville, Dayton, Houston, and New Mexico). While both Arizona and Rochester are relatively older, with the University of Rochester's programs starting as early as 1929, the others began in the 1980s. The latter programs, as evidenced by their names and emphases, have more emphasis on the "applied" part of this interdisciplinary field.

In Type B institutions, the number of faculty doing active teaching or research in optics is still very small. A typical number of such faculty members in each school are between 3 and 10. Table 3 lists, for example, Type B institutions from Ohio and its bordering states (Indiana, Kentucky, Michigan, and Pennsylvania). About 20% of the Type B institutions nation-wide have organized umbrella organizations, commonly referred to as either Centers, Institutes, or Laboratories, by combining the resources of few of its separate units [5]. The goal is to (a) coordinate and focus their activities in one or more selected areas of optics; and (b) prepare the bridge to eventually make transition to becoming a Type A institution. In fact, the majority of the Type A institutions have found it necessary first to establish such Centers or Institutes before taking their current form. In all of those instances, the Centers include at least the basic departments (physics and EE) but in many cases, participating faculty members have also come from chemistry, computer science, manufacturing, material science/engineering, mechanical engineering, and medicine.

Table 2. Type A institutions.

Institution	Type	Degree	Degree Name	Administration
U. Alabama-Huntsville	Public	MS, PhD	Optical Sc. & Engr.	Physics & EE
U. Arizona	Public	MS, PhD	Optical Science	Stand-alone
U. Central Florida	Public	MS, PhD	Electro-optics	Physics & EE
U. Dayton	Private	MS, PhD	Electro-optics	Stand-alone
U. Houston	Public	MS	Electro-optics	EE
U. New Mexico	Public	PhD	Optical Science	Physics
U. Rochester	Private	MS, PhD	Optics	Stand-alone
Rose-Hulman Inst.	Private	MS	Applied Optics	Physics

Table 3. Type B institutions.

Institution	State	Department	Optics Faculty
Air Force Institute of Technology	OH	EE	1
Air Force Institute of Technology	OH	Physics	7
Bryn Mawr College	PA	Physics	2
Carnegie-Mellon University	PA	EE	5
Cleveland State University	OH	Physics	6
Drexel University	PA	EE	5
John Carroll University	OH	Physics	4
Ohio State University	OH	EE	3
Purdue University	IN	EE	4
University of Louisville	KY	Physics	1
University of Michigan	MI	EE	11
Wayne State University	MI	Physics	9

The transition model for the formation of Type A institutions is illustrated in Figure 1. It is important to realize that as optics is changing its form from a pure science entity to one that is also applied. Accordingly, the institutions with interest in optics education are undergoing one of the following transitions: (a) Type C institutions are becoming Type B institutions; (b) Type B institutions and a few of the Type C institutions are formalizing Centers or an Institutes; and (c) Centers/Institutes are undergoing changes thus contributing to the beginning of Type A institutions. Typically, up until an institution has become a Type A institution, the primary objective of the participating faculty members remains limited to only collaborative research. Type A institutions, go a step further also to deliver a collaborative degree program.

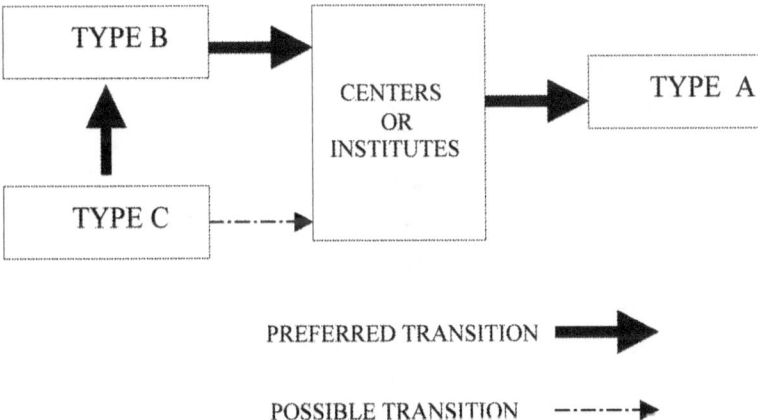

Figure 1. Transitions leading to the realization of a Type A institution.

In the specific case of the University Dayton, the optics-type faculty members were present not only within the traditional departments of physics and electrical engineering but also within the University's Research Institute. These three groups were brought into a structure of a Center that transformed eventually into a Type A program. Those who had tenure in either physics or electrical engineering continued to have tenure within their respective home units. The interdisciplinary program operated through a strategy of buying out those faculty members for the time that it needed them to teach electro-optics courses. In the long run, such an interdisciplinary program ended up hiring new faculty members who could be housed within one of the constituent departments or in the interdisciplinary program itself. In the latter case, those new faculty members, if they were to be hired on tenure-track, would need to be considered for tenure only within the interdisciplinary program.

The graduate students who entered the University of Dayton's EO Program have the following average profile: 75% have backgrounds in physics, 20% have backgrounds in electrical engineering, and 5% in other areas of science and engineering. More physics students are typically interested exploring this interdisciplinary field since it allows for more marketability upon graduation. Nearly 80% of the students come from out of State of Ohio–highest such percentage amongst all graduate programs at the University of Dayton. This number is reflective of the facts that there are only a few such programs in the nation and that the Dayton program indeed has both an appeal and a market. The entering students appear to be much better prepared as well. The average entrance GPA of the students is 3.3 for the MS students and 3.59 for the Ph.D. students. To accommodate the diverse background of its students, however, the required core of the MS program ended up being rather broad (70% of the total course work). Presence of such a large core understandably does not allow for enough flexibility, yet the Program succeeded in building marketable skills among its students through the utilization of three *integrative* laboratory courses. Indeed, in just 10 years, its students co-authored 88 technical papers that appeared in archival journals and generated over 100 papers that were presented at national and international conferences.

In Computer Engineering

Traditionally, computer science as an academic unit started out in the Sciences part of the institution and utilized until recently faculty members with terminal degrees in either mathematics or physics. As many of these departments began to produce doctoral graduates in computer science, the next generation of computer science faculty members ended up having doctoral degrees in computer science rather than physics or mathematics. In part due to this make-up, computer science as a discipline limited its scope to software and software related issues. However, starting early 1980's more electrical engineering departments began to take more interest in not only hardware but also hardware-software integration issues. This latest twist of technological growth and demands of the market thus called for a new breed of engineer, namely, computer engineers, somewhat distinct in training and skills from traditional computer science and electrical engineering graduates.

As identified by NSF, using Corporate Technology Information Services data, "about half of the new high-tech businesses formed during the past two decades were computer related companies" [1]. The percentages of new high-tech companies formed during the periods 1970-89 in the four most active areas is shown listed in Table 4. The top-most two active areas involve computing systems. High-tech companies formed during the 1980s showed their importance to the US economy by their performance as manifested by four indicators-employment growth, job creation, annual sales, and sales exports. According to NSF as shown in Table 1, "computer-related companies experienced the highest growth rate of the seven technology fields, increasing employment by about 14% and adding over 250,000 new jobs" [1].

Table 4. Growth in the number of High-Tech Companies in 1970-89 [1].

Field	1970-74	1975-79	1980-84	1985-89
Automation	11%	10%	9%	7.5%
Biotechnology	3%	4%	5%	6%
Hardware	16%	19%	20%	18%
Software	23%	37%	35%	23%

According to the Engineering Workforce Commission of the American Association of Engineering Societies, "sustained growth in the numbers of computer specialists propelled that occupational group well beyond any of the more traditional engineering disciplines in its sheer size" [6]. For a while, the largest job growth is not necessarily in the area of traditional electrical engineering discipline but in that which is a hybrid of both electrical engineering and computer engineering. Prospective students and their parents interested in electrical engineering, who are often also interested in dual majors or availability of electives, are beginning to decide on the institution based on their perception as to whether or not a computer engineering program is also available.

A typical computer-engineering curriculum thus has an overlap of both computer science and electrical engineering [7]. This overlap can be explained using Table 5. As one goes to a column on the left, the curriculum becomes more physics-oriented, while a shift to the right corresponds to emphasis on computer-related topics. A typical electrical engineering curriculum restricts itself to including subject matter from Columns A through D while that of computer engineering involves subject matter from Columns B through E. We need to understand that Column A is often viewed as an overlap between physics and electrical engineering. Depending on the history of an institution, a computer science department may end up invading a part of Column D while an electrical and/or computer-engineering department may have already invaded a part of Column E.

Table 5. Traditional Electrical and Computer Engineering Curriculum.

A	B	C	D	E
Applied Physics	Signals and Systems	Circuits	Hardware	Software
Electromagnetics	Signals	Analog	Logic Design	Programming
Energy	Linear Systems	Digital	Architecture	Data Structures
Machines	Control	IC Design	Networks	Software Engineering
Solid State Devices	Communication	VLSI Design	CAD	Operating Systems
Optics	Signal Processing		Concurrency	

The evolution of computer engineering degree programs thus involves using of expertise from both computer science and electrical engineering. Based on an extensive survey conducted of all computer engineering programs in the US in September 1995, the overwhelming majority of computer engineering programs (65.6%) makes use of software courses offered by the CS Departments. To meet the computer engineering challenge, accordingly, different institutions have gone through one of the following structural transformations:

(a) *A stand-alone computer engineering (CE) department is created within Engineering* by hiring either computer engineers and/or electrical engineers. In this process, some of the faculty members end up moving out of the traditional electrical engineering department. This route is pursued when computer science, already housed within the Sciences, decides to remain within the Sciences. Usually, it is more expensive to initiate a stand-alone CE unit. Most of these departments were formed in the mid-1980's when many traditional electrical engineers failed to see engineering value of this new discipline. However, now more than ever before an electrical and computer engineering department rather than a computer-engineering department is formed. This transition can be described as: EE→CE+EE, i.e., where a separate EE department also continues to exist along with this new computer engineering unit.

(b) *A computer science and engineering (CSE) department is created always within Engineering* by combining faculty members from computer sciences and those electrical engineers who are closer to Column D topics. Such academic departments typically offer degrees in both computer science and computer engineering. In some of these cases, computer science department may have already moved within the Engineering long before the formation of this new unit. This transition can be described as: CS+EE→CSE+EE, i.e., where a separate EE department continues to exist along with this new CSE unit.

(c) *An electrical engineering and computer science (CSEE) department is formed in Engineering* by combining both electrical engineering and computers science faculty members. Typically, these departments offer two degrees-computer science and computer engineering. Often computer engineering is only a concentration or specialization in one or both of those two degree programs. This transition can be described as: CS+EE→CSEE, i.e., where neither a separate EE nor a CS department exists along with this new CSEE unit.

(d) *An electrical and computer engineering and computer science (ECECS) department is formed in Engineering* by combining (i) electrical and computer engineering (ECE) and computer science faculty members, or (ii) electrical engineering (EE) and computer science faculty members. Again, typically, these new departments offer three degrees-electrical engineering, computer science, and computer engineering. This transition can be described by either CS+EE→ECECS, or ECE+CS→ECECS, i.e., where neither a separate EE department nor a separate CS department exists along with this new ECECS unit.

Transformation (as in d above) results in the creation of a mega-department within Engineering. Some of the more recent examples of such institutions include Case Western Reserve, Cincinnati, Ohio University, and West Virginia. Table 6 shows a breakdown of nation-wide engineering schools and colleges as listed in ASEE data book [8]. The data compares the change that took place between 1991 and 1995 in terms of the number of institutions having (a) a stand-alone EE department; (b) no EE department; (c) either electrical and computer engineering (ECE) or electrical engineering and computer science (EECS) department; and (d) neither EE nor Computer-related (C-R) departments. The increase in the number of institutions having engineering programs is due to the addition of relatively smaller institutions that, in most cases, have only a single engineering department. To a certain extent, this increase is equivalent to the increase in the number of institutions having neither EE nor C-R departments. We notice though that the number of stand-alone EE departments has decreased when simultaneously the number of either ECE or EECS departments has increased. This suggests that in many of the institutions, EE departments have ended up becoming either ECE or EECS departments. A majority of these ECE departments offers a computer engineering degree program in addition to an undergraduate electrical engineering degree program while the rest offer at least a computer engineering concentration.

Table 6. Make-up of the Engineering Schools/Colleges listed in ASEE data book.

Measure	Number (1991)	% (1991)	Number (1995)	% (1995)
Stand-alone EE Dept.	135	60	119	51
No EE Dept.	1	0.4	1	0.4
ECE or EECS Dept.	73	32	91	39
Neither EE nor C-R Dept.	16	7	23	10
Total	225	100	234	100

It may be worth exploring the nature of evolving C-R departments within Engineering as also listed in the ASEE data book [8]. Table 7 breaks down the number of institutions (having computer-related programs in Engineering but where, at the same time, there isn't a stand-alone EE department) as to how computer-related programs are configured. We see that the number of engineering units that deals with computer-related programs have increased from 138 to 166 in just four years. The numbers of ECE, CS, CSE, and ECECS departments have all increased with most growth shown in ECECS (electrical and computer engineering and computer science) category. The stand-alone computer engineering (CE) departments are vanishing because they are being merged with other suitable entities. The number of computer information system (CIS) departments having a certain business flavor remains unchanged.

Table 7. Computer-related program configurations in Engineering.

Type	Number (1991)	% (1991)	Number (1995)	% (1995)
ECE	67	49	73	44
CS	35	25	40	24
CE	9	7	7	4
CSE	18	13	25	15
ECECS	6	4	18	11
CIS	3	2	3	2
Total	138	100	166	100

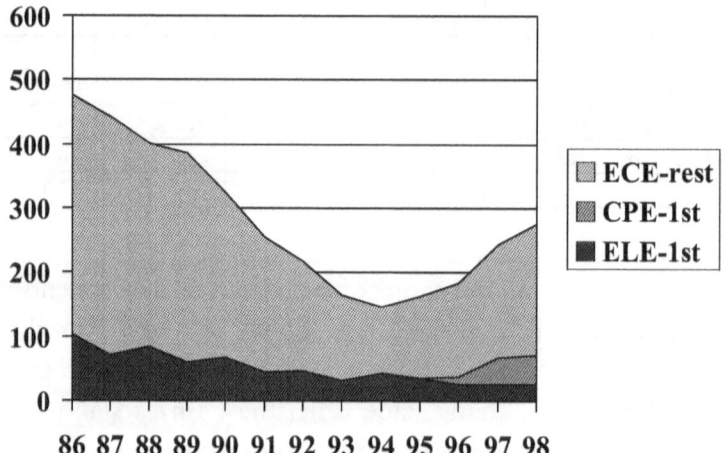

Figure 2. The number of freshman computer engineering (CPE-1st) and electrical engineering (ELE-1st) students are plotted along with the number of remaining undergraduate (sophomore, junior, senior, and transfer) electrical and/or computer engineering students (ECE-rest) at the University of Dayton.

Figure 2 shows the enrollment numbers for the freshman class of both computer engineering (CPE-1st) and electrical engineering (ELE-1st) at the University of Dayton. The total enrollment that includes ECE-rest (sophomore, junior, transfer, and senior) for the Department shows a continual decline that reached its minimum in 1994. With a newly-started computer engineering program that enrolled its first batch of students in 1996, the total enrollment of the Department is picking back up with 1998 number reaching a value that was similar to that in 1991. Note that the ELE-1st numbers have not improved and, thus, it may be concluded that the most recent growth in the numbers of total undergraduate students is directly related to the initiation of a computer engineering program. This trend is becoming also evident throughout the nation. For example, consider Table 8 that lists the number of undergraduate students at California State University, Sacramento, which initiated a Computer Engineering (CPE) program over a decade ago. We observe a continual but steady shift of students from Electrical Engineering (ELE) to CPE. This trend is generally true even in the institutions that have added computer engineering programs only recently.

Table 8. Enrollment Trends in California State University, Sacramento

	F90	F91	F92	F93	F94	F95	F96	F97
CPE	150	174	188	212	220	237	269	325
ELE	536	445	425	419	395	340	375	385
Total	686	619	613	631	615	577	644	710

(Source: *http://inst-srv1.adm.csus.edu/ECS/ECS.TOC*)

Where Do We Go From Here?
The trends have been that interdisciplinary Electro-Optics programs have succeeded in only a few universities. In the majority of those cases,

institutions wanted to cultivate niche areas so that they could capitalize on the resources that were already available regionally within either the industrial sector or government laboratories. The computer-engineering field, in comparison, is making dramatic inroads throughout the U.S. institutions of higher learning. The following trends are becoming clear: (a) computer science departments are moving out of the Sciences and into Engineering; (b) electrical engineering departments are accommodating the societal need to offer computer engineering degree program either alone or by joining with computer science; and (c) without a computer engineering program in place, electrical engineering departments will continue to have declining undergraduate enrollments.

The above two examples of interdisciplinary programs demonstrate Rustum Roy's seminal argument made some twenty-two years ago that "the real problems of society do not come in discipline shaped blocks...All societal problems are necessarily complex, and a mixture of so-called disciplines is always needed for their study....In order to be able to actually study and do research on societal problems, therefore either an interdisciplinary or multidisciplinary mix is mandated" [10]. And so new disciplines come into existence from either fusion or fission as a result of interdisciplinary or multidisciplinary activity. Lewis Thomas says that "The endeavor (science) is not, as is sometimes thought, a way of building a solid, indestructible body of immutable truth, fact laid precisely upon fact in the manner of twigs in an anthill. Science is not like this at all: it keeps changing, shifting, revising, discovering what is wrong and then heaving itself explosively apart to redesign everything" [11]. Similarly, disciplines change, often in interdisciplinary or multidisciplinary activity. In slightly different terms, Ernest Boyer and Fred Hechinger describe the always on-going process of scholarly investigation, interdisciplinary or otherwise, and its discoveries in this way: "The process of scholarly investigation never ends. Each new discovery poses new problems, opens up new options, and reminds us in a fascinating and frustrating way, that with all of our supposed wisdom, we are only learning how to learn" [12].

What, then, is needed to develop interdisciplinary academic programs? One answer is timely administrative and policy changes, however incremental, in universities and colleges to support the compelling changes that must be addressed if we are to provide meaningful academic programs evolving at the interconnections of disciplines in response to rapidly changing needs in society. Rustum Roy makes this point very persuasively in his paper on "Interdisciplinary Science on Campus–the Elusive Dream" which is a bold discussion of the central issues in interdisciplinary activity some twenty-two years later that, "the problems of society always and necessarily involve many disciplines in their study and solutions…The primary reason for this is that the best research can be done only in the context of the whole problem" [12]. But the difference in making that point again, here, is that the demand for change today is much more rapid than it ever was, so that interdisciplinary programs evolving in this dynamic environment have *increased* rather than *decreased* the number of academic specializations available to graduate students. To be sure, the changes do not take place at the same rate across disciplines on a campus. Indeed, changes in disciplines that are in constant demand in society are very rapid, in others perhaps not as rapid. Even where disciplines have coalesced into new disciplines as in the example of cognitive science, the convergence of the disciplines of biology, psychology, electrical engineering, psychology, and philosophy resulted in new specializations, not fewer.

The present demand for change is continually fuelled by the demand for new competencies resulting from an almost boundless access to information and from knowledge discoveries supported by such access, particularly in science and engineering disciplines. Students develop competencies in disciplinary interactions where they discover new knowledge, develop a skill, and develop increasing intellectual confidence. This is not to say that competencies can be developed only in interactions between disciplines. It is to assert that the fissioning and fusion of disciplines, as illustrated in the two specific instances we give earlier, provides an expanding scope of specializations and new interfaces between disciplines that can sharply focus the skills students could acquire in response to the demand for new competencies.

At many universities and colleges, interdisciplinary programs which bridge emerging technologies in a logical way have become part of the research culture–particularly at universities with academic medical centers and/or engineering schools/colleges. Such programs often involve emerging interdisciplinary areas and are competing for large grants involving multiple investigators. Inevitably, because of their scope, such programs involve investigators from different disciplinary areas. The investigators cross disciplinary boundaries to bring to bear a host of techniques, many bodies of knowledge, and a variety of supporting staff, all engaged in studies designed to address broad based problems in human health or advanced technology. These problems usually require an interdisciplinary approach because, as Roy remarks, real problems do not come in discipline shaped blocks.

What, then, are the changes we need in the way we educate students engaged in interdisciplinary projects? As far as training students, here is an opportunity to develop a program for students designed specifically for those working on team-driven interdisciplinary projects. The goal is to design a procedure for such a training program, and to seek support for planning and implementation from foundations and companies, especially for curricular planning that supports the new training program and support students in the training. As identified earlier, the more viable electro-optics programs are all co-located with either major government laboratories or high-tech industries. The electro-optics interdisciplinary efforts, therefore, invariably benefit from having a better chance at incorporated the research problems of these co-located and other sponsors. By their very nature, the majority of their problems are always multidisciplinary as well as developmental which in turn, thus, provides for a serious interdisciplinary efforts. The typical electro-optics problems thus involve electrical engineers, physicists, material scientists, human factor specialists, mechanical engineers, biomedical engineers, and marketing. It is very well established that a typical computer engineering curriculum will always invariably involve mathematicians, electrical engineers, and computer scientists. A team-driven interdisciplinary problem solving in disease-related research involves basic scientists, medical researchers, clinical researchers, social workers, psychologists, epidemiologists, nutritionists, medical anthropologists, and ethicists among others. Such a global approach has multiple goals and seeks for instance a molecular basis for disease, a rationale for development of treatment, treatment testing and analysis, epidemiology/public health, compliance, behavior modifications, and the like. In team-driven interdisciplinary problem solving in electro-optics or biological research, for example, opportunities clearly exist to develop a unique approach to training students in new fields fissioned in demand to rapid changes in society.

What changes do we need in the ways in which student contributions to such projects are supported and evaluated? Here again, to adapt to the reality of a different training environment, we need a new curricular environment that supports and nurtures the student in training. We need to analyze the skill set required for participating in and leading an interdisciplinary project. Then, we need to determine how to educate the students in the teams in those skills. These can include broader background courses that could define research problems in the context of long-term goals, and learning about people skills to make the best use of group dynamics. Policy issues include sharing credit, allocating contributions and resources, changing the criteria for measuring academic success for students. Other issues equally important involve questions such as: should students coming from different backgrounds on a team be viewed as a cohort or not ? What traditional values and elements are affected in a common training program across disciplines? What should be the goal of such an approach–should it, for example, be expertise in a specific area but understanding of global issues, or knowing the extent to which approaches are used by other team members, or developing a clear sense of one's own role and contributions, or all of the above ? How will the design of complex engineering systems relate to its environmental impact, reproducibility, reliability, efficiency, cost, safety, by-products, and ethical issues?

The multidisciplinary component of engineering programs is also expected to be on the rise. The Accreditation Board of Engineering and Technology's Criteria 2000, for example, is requiring engineering programs to prepare their graduates with skills to function on multidisciplinary teams, and a broad education to understand the impact of engineering solutions in societal and global contexts. There are clear signs already that programs are initiating efforts to integrate suitable bridges primarily with other engineering disciplines and topics. It can be anticipated that there will be other collaboration with sciences, business, and perhaps with humanities disciplines. Indeed, there is already a recognition that electrical engineering majors, for example, may be better off being exposed to also biology as opposed to only physics and chemistry.

It should be recognized that no interdisciplinary efforts have a chance of survival without the efforts of its key stakeholders–committed interdisciplinary faculty, an enthusiastic champion, and a very sympathetic administration. These commitments are particularly important since interdisciplinary efforts are viewed by majority of the university faculty both as a distraction and a resource drain.

What are the institutional changes we need for making successful transitions to the interdisciplinary training environment? Support from the academic leadership is essential because the new environment crosses traditional separations of areas–a single program could have participation from engineering, mathematics, social sciences and biomedical sciences where each traditional discipline may reside in a different school or college within the university. Barriers abound in such situations–reluctance to share or divide tuition, resistance to faculty teaching outside the school or department, resistance and uncertainty concerning the evaluation of the impact of faculty work in evolving interdisciplinary areas, concern by students about their professional identity, and the like. Affirmation is required at the level of President and Provost that involvement in such activities as an institutional priority is a requirement for overcoming these barriers. Buy-in by deans and department chairs/heads is also required, and the process is, of course, catalyzed by selective allocation of resources.

Policies that provide incentives for cooperation are thus essential. Opportunities to support students on interdisciplinary training grants can be a force for change at the departmental level. Specifying special recognition of team-based interdisciplinary activity in promotion and tenure decisions, and in annual (salary) reviews can provide strong incentives for faculty involvement. Administrative facilitation of cost and revenue sharing between schools and colleges involved in interdisciplinary programs can help lower barriers to involvement in these programs. Activation barriers to initiating new programs can be lowered by tangible support of new initiative, e.g., by small grants for pilot programs and/or interdisciplinary course development. These efforts, it is hoped, will catalyze the movement to this new mode of research-based education. The employers' need for interdisciplinary-trained student and grant support for truly integrated interdisciplinary research programs will provide the incentives to continue such programs once they have been initiated.

Notes

1. *Science & Engineering Indicators*, 10th Edition, National Science Board, National Science Foundation, Washington, DC., 1991.
2. *Engineering Education for a Changing World*, American Society of Engineering Education, 1994.

3. K. Cupery, "Optics education: supply and demand," SPIE, Vol. 978, pp. 116-120, 1988.

4. *Guide to Optics Courses and Programs in North American Colleges and Universities 1990*, Master's and Doctoral Level edition, Optical Society of America, Washington DC, 1990.

5. *Optics Education*, SPIE-The International Society for Optical Engineering, Bellingham, WA, 1991.

6. *Engineers*, Engineering Workforce Commission, American Association of Engineering Societies, Washington, DC, January 1995.

7. *Electrical and Computer Engineering at Carnegie Mellon: A New Curriculum*, Carnegie Mellon University, 1995.

8. *ASEE Profiles of Engineering & Engineering Technology Colleges, 1995-96 Academic Year*, American Society of Engineering Education, Washington D.C., 1997.

9. Rustum Roy, "Interdisciplinary science on campus—the elusive dream," *Chemical and Engineering News*, August 29, 1977, p.29.

10. Rustum Roy, "Interdisciplinary science on campus—the elusive dream," *Chemical and Engineering News*, August 29, 1977, p.29.

11. Lewis Thomas, "On the Uncertainty of Science," Harvard Magazine, Vol. 83, Sept-Oct 1980, p. 20.

12. Rustum Roy, "Interdisciplinary science on campus—the elusive dream," *Chemical and Engineering News*, August 29, 1977, p.29.

Experience With I³R in Materials Research: Worldwide Programs

Early Stirrings of
Materials Science in Britain ▶▶

Robert W. Cahn
Department of Materials Science & Metallurgy
University of Cambridge

I was educated in 'Natural Sciences' at Cambridge University during the War, specializing in metallurgy. The program also included a first class course entitled "mineralogy," which in fact focused on crystallography, x-ray diffraction, crystal physics and crystal chemistry. That was quite my favorite course and, as I can see in retrospect, whetted my appetite for a broader approach than was permitted by classical metallurgy. The metallurgy course in fact was rather boring, illuminated only by a few lectures on the Bragg-Williams theory of order-disorder transformations, which were a revelation. That, in turn, inclined me towards research on intermetallic compounds, years later. When I started out on doctoral research, at the Cavendih Laboratory, the delightful professor of metallurgy, Robert Hutton, lent me his precious German copy of Schmid and Boas' book "Kristallplastizitaet," which I learnt more or less by heart. That had a big effect also and persuaded me to accept my first research assignment after the doctorate, on twinning in metallic uranium.

The foregoing is intended to underline how one's youthful intellectual experiences shape one's concerns and obsessions. The variegated influences I have listed predisposed me, I believe, to respond to the ideal of materials science when that came along in the late 1950s, so much so that at present I am engaged in writing a history of materials *science*.

For 11 years, 1951-1962, I taught and researched at Birmingham University, in the Department of Physical Metallurgy. There, the department chairman, Daniel Hanson, a metallurgist who in his youth had been right hand man to Walter Rosenhain, the inventor of physical metallurgy, had put in place an entirely new way of teaching metallurgy, with a strong emphasis on physical fundamentals: Alan Cottrell and Geoffrey Raynor, youthful professors, put into effect the new approach, which certainly removed the yawn-inducing character of the older methods. The Birmingham course, uniquely in its time, moved halfway towards modern ideas of materials science: attention was still focused on metals, but the new approach to metals cleared the way to the materials science courses such as that pioneered at Northwestern University after 1958. Cottrell's little textbook of 1948 [1] was a harbinger of spring. The innovations in Birmingham in teaching were matched by the new research approach instituted by Cyril Stanley Smith at the Institute for the Study of Metals at the University of Chicago, 1946-1961, until that institute was hijacked by physicists and chemists at the University. Another major development which fostered the approach pioneered in Birmingham and Chicago was the creation of Acta Metallurgica in 1953; I had a paper in the first issue and have favored it ever since as the best journal of its kind, anywhere.

When, in 1962, I was offered the first chair in Britain with 'materials' in its title—the chair of materials technology at the University College of North Wales, in Bangor—I jumped at the chance. It turned out, however, in spite of the broad title, that I was expected to focus exclusively on semi-conducting materials, and so I moved again 2 years later, to the recently founded University of Sussex, where on the first day of 1965 I became professor of materials science (the first chair with that title in Britain). Almost at once, presumptuously, I published my view of what materials science was [2], together with an optimistic overview, "Materials of the Future" [3]. Given the unusual structure of the new university, I was able to teach my new discipline in 3 distinct ways: (a) an engineering context,

(b) a chemical context, and (c) a physical context. It was a fascinating experience to set all this up and to institute a very broad range of research as well. By 1969 I was ready to publish an account of how the subject was taught at Sussex University [4]. I remained in charge of materials science at Sussex for 17 years, until in the middle of a financial crisis (British universities have had more than their due share of these) it was killed off by jealous mechanical and electrical engineers.

While I was still at Bangor, in 1962-3, I learnt of moves to set up a Materials Science Club, and joined at once. This Club was the brainchild of a chemical engineer working in the oil industry, Leslie Holliday, and it became a very successful though low-key group that flourished through the 1960s, 70s and much of the 80s before it was swallowed up in the Institute of Materials in London and disappeared. It published a quarterly Bulletin of which I still have a few issues, up to #59 in 1979. I also still have a "Who's Who" of members in 1979, some 450 in all, from a wild mixture of educational establishments (at many levels), industrial firms and many government laboratories in Britain, even a journalist or two. The Club brought together all the disciplines as American MRLs at about the same time, and it organized a splendid series of symposia (MRS-style), but never in parallel sessions. A few examples: The mechanical properties of biological materials; crystallography in materials science; the safe use of structural materials in building; standards and specifications as an aid to safety. In the Bulletin of September 1978, I gave an account of the infant Materials Research Society, having not long before attended my first meeting of that body. At that time, the MRS in America and the Materials Science Club in Britain were both run by volunteers on a shoestring; the MRS moved on from there to much more elaborate and professional arrangements; the British body never did, but it was nevertheless an important milestone in the development of the discipline in Britain.

By the time I left Sussex University, in 1981, quite a few British professors had made the paradigmatic shift from metallurgy to materials science, though the main professional body, the Institute of Metals, was at that time still deeply resistant to the new concepts. A particularly important development, in my view, was the Open University. This remarkable university concentrates on "distance learning" for mature students who cannot attend a normal university, because they are in fulltime employment, or married with children, or (most commonly) do not have formal school-leaving qualifications. The founding fathers of the OU laid it down that "we take it as axiomatic that no formal academic qualifications would be required for registration, and only failure to progress adequately would be a bar to continuation of studies," and this principle has been adhered to for 30 years. Many of the students were dissuaded from trying to learn by uninspiring high (secondary) schools and only acquired an appetite for study in later life. Some 5000 students graduate every year. Study involves the use of purpose-written printed study modules (special, distinguished products of the Open University system), frequent TV programs broadcast by the BBC at times when other demand for TV is at a minimum, regular face-to-face meetings with regional tutors, and an annual week of seminars or laboratory work at friendly 'regular' universities. Students also take written examinations, both multichoice questions, regular problems and essays. Materials have been a field of study almost from the beginning. I was involved as a professional adviser and examiner in the early stages. The 'course modules' each deal with a limited field of MSE (e.g., semiconductors and devices), they are splendid teaching documents, prepared by expert teams, and are not as well known as they deserve to be though they can be bought from the Open University (*http://www.open.ac.uk*). Nowadays, there are a number of other distance-learning institutions that encompass materials science, but the OU was the first.

Another route by which I was enabled to contribute to the development of materials science in Britain was opened up in 1967, when John Maddox, the editor of *Nature*, invited me to become materials science correspondent to the journal; in the 32 years since then, I have published about 100 essays on many topics in materials science…popularization of science for scientists. Many of these (up to 1992) were collected in a book [5]; some 15 more articles have appeared since that book was published. I can recommend this activity as an excellent way of learning about materials science!

Some other broad-based essays which deal with materials science and engineering in abroad way include an essay-review of the COSMAT report [6], an article on case-histories of innovation (mostly of novel materials) [7], one on the relation between solid-state physics and metallurgy [8] and one drawing together many aspects of structural disorder [9]. Finally, I should mention a long chapter on the history of physics of materials (it could just as well have been entitled 'materials science') in a recently published work on the history of 20th-century physics [10]. This was a 'dry run' for *The Coming of Materials Science,* a book now in preparation.

[1] Cottrell, A.H., *Theoretical Structural Metallurgy* (Edward Arnold, London, 1953).

[2] Cahn, R.W., *What is Materials Science?*, Discovery (July 1965).

[3] Cahn, R.W., *Materials of the Future,* The Technologist (1965) 163-172.

[4] Cahn, R.W., *Materials Science Survey: Teaching Materials Science at Sussex University,* Materials Research Bulletin Vol. 4 (1969) 699-704.

[5] Cahn, R.W., *Aritifice and Artefacts* (Institute of Physics Publishing, Bristol and Philadelphia, 1992).

[6] Cahn, R.W., *Materials and Man's Needs (review of the COSMAT Report),* Technology and Culture (1975) 570-575.

[7] Cahn, R.W., *Case Histories of Innovations,* Nature Vol. 225 (1970) 693-695.

[8] Cahn, R.W., *Solid State Physics and Metallurgy,* Physics Bulletin (journal of the Institute of Physics, London) Vol. 36 (1985) 207.

[9] Cahn, R.W. and Johnson, W.L., *Review: The Nucleation of Disorder,* Journal of Materials Research Vol. 1 (1986) 724-732.

[10] Cahn, R.W., *Physics of Materials, in Twentieth Century Physics* (Edited by L.M. Brown, A. Pais and Sir Brian Pippard), (Institute of Physics Publishing and American Institute of Physics, 1995) Vol. 3, 1505-1564.

I³R—A Personal Canadian Experience ▶▶

Patrick S. Nicholson

Department of Materials Science & Engineering
McMaster University

Canada is the second largest country in the world yet its population is~30 million, the same as metropolitan Tokyo! The people spread the border like weather-stripping and universities and industries are concentrated therein. The major universities with efforts in Materials are University of British Columbia, the University of Calgary, the University of Manitoba, the University of Toronto, McMaster, Queens, McGill, Ecole Polytechnique, Laval and Dahousie. From the viewpoint of I³R, British Columbia, Toronto and McMaster had formal efforts therein. This is a personal view of I³R at McMaster University, Hamilton, and Ontario.

I arrived in Canada in 1969. Newly formed at McMaster was the "Institute of Materials Research" an association of academics, scientists and engineers on campus, with interest in materials. The Dean of Science, Howard Petch had approached the National Research Council in Ottawa to fund faculty, technicians and equipment for a material research enterprise and as a result the Council cast a new program termed Negotiated Development Grants. The first of these were awarded to McMaster and the Universities of Toronto and British Columbia. The McMaster Institute is the only one to survive. One possible reason for this vitality is investment in capital equipment and professionals rather than the latter exclusively—a definite real reason is the first Director, Dr. James A. Morrison. Jim was Director of the Institute for Materials Research at McMaster from July 1969 to 1986. Jim Morrison understood that if Chemists, Physicists, Geologists, Civil

Engineers, Mechanical Engineers, Chemical Engineers, Biochemists, Engineering Physicists and Pathologists were to research together, they must laugh and drink together. His appreciation of this was his secret. Scientists and engineers of all stripes take coffee (or tea), 10:00-10:30 and 3:00-3:30 each day so one can depend on seeing a colleague. Here a voice asks, "How do you stop the chemists, physicists and others sitting by themselves?" This was the genius of Morrison. He arranged seminars, with donuts and coffee—evening gatherings, with "refreshments"—and he organized cross-fertilization in uberous milieux! How scientifically effective was this conviviality? The number of co-authored publications crossing department lines, 1983-86, were 116*! Today's Institute had a 1998 budget of $1 million. There are 8 dedicated technical staff servicing x-ray, electron microscope and crystal-growth facilities. There are 3 office staff. Faculty members occupy university offices and laboratories so the Institute occupies minimum space.

Another factor that ensured dialogue between scientists and engineers is the Natural Sciences and Engineering Research Council (NSERC). Formed from the NRC in 1969, it exclusively funds university research. The NSERC Budget for 1999 was $477.9 million. The NSERC program of interest to I³R is the Strategic Grants. This program was specifically designed to promote university research in areas of strategic importance to Canada. The program was instituted in 1977-78 and presently identifies the following strategic areas; Biotechnologies, Information Technologies, Energy Efficiency Technologies, Environmental Technologies, Manufacturing and Processing Technologies and Materials Technologies. The latter area impacts I³R in Canada.

The present Committee of Materials Technologies consists of 5 university Materials Faculty from Canada, two from the US and representatives from the following companies; In NOVA Corp, IPSCO, International Paper Co. and Woodhead Consulting. External referees are involved in proposal assessment. A committee member serves for 3 years. Two million dollars were dispersed by the committee in 1998, the average award being $106,000. This total involves ZERO overhead. The Provinces using monies from the Federal Government fund education in Canada. The Province of Ontario, Research Leadership Fund, awards universities 12% of the total Federal Council Grants they are awarded, (averaged over the past three years). This is for overhead application.

Critical to the success of Strategic Grant application is *real* Canadian industry interest. The latter does not involve cash contributions but strong supporting letter, firm details of involvement and in-kind assistance. It is this aspect of the program that further "persuades," chemists, physicists and other faculty members in pure science to dialogue with faculty members of the Institute with a practical bent. Thus public policy catalyses I³R and makes it work for the Institute of Materials Research at McMaster.

*Dates chosen to avoid the High T_c superconductor phenomenon.

Priorities in R & D of Materials for the Next Millennium ▶▶

Hiroaki Yanagida
Research Institute
Japan Fine Ceramics Center

Significance Of Entering Next Millennium

We discuss the future based upon knowledge about the past. Shortly we are entering next millennium. It is especially challenges to discuss the future considering a term of thousand years.

The Ceramic Society of Japan and The American Ceramic Society recently celebrated their centennial anniversaries. Establishment of the professional societies means birth and unification of ceramists. Professional groups like ceramists, alchemists and scientists are recognized in a term of hundred years. The term of hundred years means the birth of professional groups with a suffix–"ists". Back about a 1000 years, we do not see such "—ists" professional groups. At least we recognize no professional ethics and logic there. The tales of Genji was written by Murasaki-shikibu around 1,000 years ago. People made necessary things for themselves. Murasaki-shikibu was not a professional novelist. The oldest earthenware ever found was excavated from Kanita-town of Aomori prefecture, Japan and it is considered only around 15 thousand years ago. No serious environmental disasters in the last millennium.

The history of civilization is the history of exploitation. Irrigation caused enrichment of salinity in soil. Rapid economic growth was made possible by rapid consumption of subterranean resources. So-called advanced technology has now made the general public alienated. Upon reconciliation of these facts we have to redesign scheme of our technology on the millennium scale.

Important issues of R & D for the next millennium are, therefore, paying attention to environmental impact, people's unwillingness to adapt to technology, recognition of the roles of the ancestors of ceramics.

Directions of Technology, Miniaturization, Enlargement, Integration and Brevity

Yanagida proposes that key directions of R & D in technology are: miniaturization, enlargement of structures, integration and pursuing brevity. It may appear that these conflict with each other. However, they are based upon common concepts.

Miniaturization of devices saves space, resources and energy. So-called advanced technology has been and still is being developed along this direction. Yanagida is directing a National Project 'Frontier Ceramics' sponsored by STA Japan. Interfaces are considered as prominent sources of novel and important functions. The project's aims include the plan that interfaces be well designed and analyzed. Especially non-linear interactions between two different materials including crystal axis orientations receive much attention. Non-linear interaction is a source of integration.

Large scale structures require much energy and resources. Development of technology related to large scale structure, therefore, is very important for saving energy and resources. Maintenance technology is also to be developed to save resources. The key-word in technologies related to large scale structures is simplification and plainness. Yanagida and his group have developed technology where strengthening and damage monitoring capability are fulfilled simultaneously by only one action. Simultaneous fulfillment of more than two requirements by one action leads to the concept of integration. Integration makes technology simpler and plainer.

Ken-materials research consortium presided by Yanagida is trying to develop plain technology which is well accepted by general public. Logo-mark is shown in Figure 1. Key words are 7 Chinese characters.

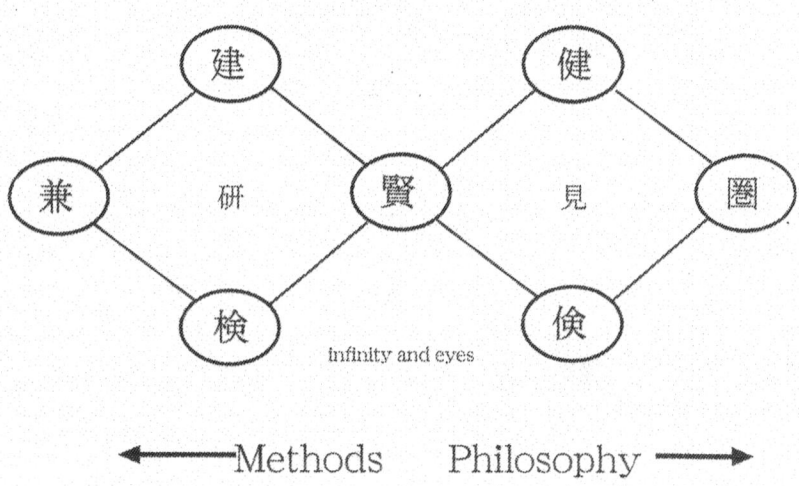

「賢」denotes wisdom of autonomy or self-control,

「建」structural reliability and performance,
「検」functional capability such as self-diagnosis or self-control,
「兼」integration of structural and functional design,
「健」soundness for general public to commit to establish Techno-democracy,
「倹」simplicity to avoid spaghetti syndrome of modern technology, and
「圏」contribution to environmental issues, respectively.

「賢材」=ken-materials
「賢材研究会」=ken-materials research consortium
「研」= research
「見」= to see

Figure 1. Key words are 7 Chinese characters

Integration of structural reliability and capability of self-damage monitoring is said to have opened new dimension in R & D of composite materials. Yanagida is recently frequently invited to International Congress related to materials design to present the concept of integration.

Yanagida has proposed a typical case for self-damage monitoring capability and improvement of mechanical performance in CFGFRP, carbon fibers and glass fibers co-reinforced plastics. Shimidzu Co.Ltd. and SOK Co.Ltd collaborated him. If we start with FRP, glass fiber reinforced plastics, an action of mixing with carbon fibers improves elastic modulus and gives rise to capability of damage monitoring by checking change in electric conductance. If we start, on the other hand, with CFRP, carbon fiber reinforced plastics, the action of mixing glass fibers improves stiffness to avoid sudden fracture. Figure 2 shows relation between load and deformation. CFGFRP has been developed as an alternative of steel bars to reinforce concrete structure to avoid troubles arising from erosion of steel. Yanagida has reached the material from a completely different view point to design strengthening and capability of self-damage monitoring simultaneously. In CFRP we can measure electric conductance. However, the loss of electrical conductance there means fatal fracture of the material. The change in electric conductance before the carbon fibers undergo fracture is very small to detect. Mixture with glass fibers makes possible the material stand further behind the point the carbon fibers suffer fracture. A remarkable change is observed while the material remains still unfractured. The decrease in conductance corresponds the portion of fractured carbon fibers. Life detection is possible by measuring the change. Latent damage after an earthquake of large scale structures such as bridges, highways, buildings can be monitored by measuring the change in conductance including carbon fibers. Damage by invaders through the shell walls of safety chamber can be delayed and monitored by applying the CFGFRP to reinforce the walls. Mixture of carbon fibers with FRP enables not only improvement of mechanical performance of the material but also monitoring of damage. No additional sensors to monitor damage are added here. This is the typical case of integration and integrated material.

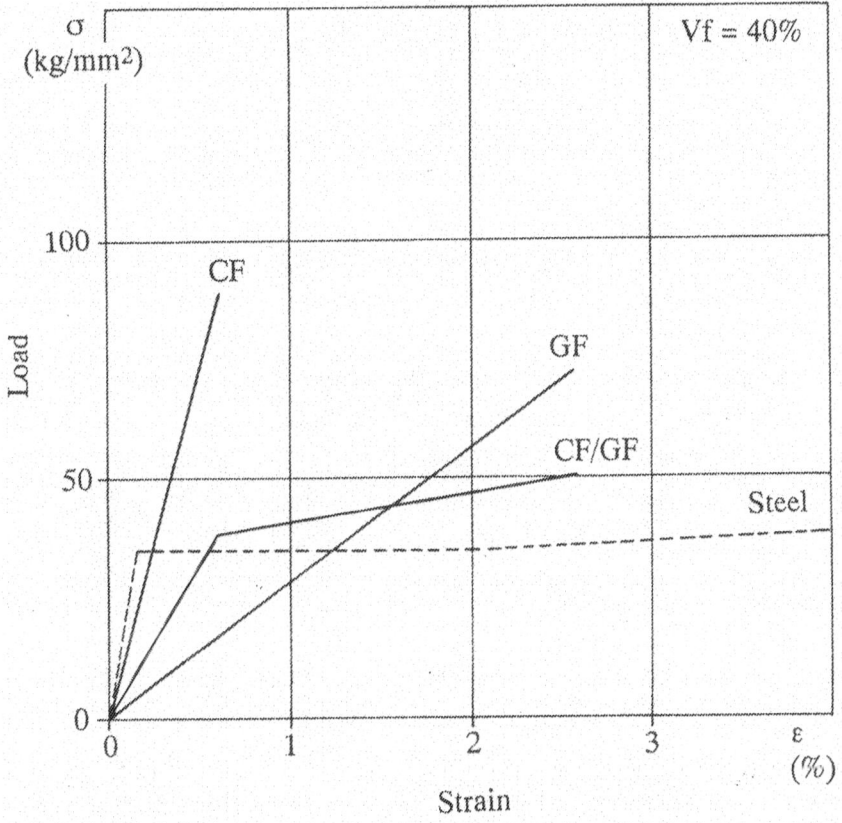

Load/Strain Relation
(Steel, Carbon Fiber, Glass Fiber, CFGF Composite)

Figure 2

Breakdown of carbon fiber usually takes place when the fiber is elongate around 1.0%. In order to check damage due to strain less than 1.0%, change in percolation of carbon powders around the fibers to reinforce ceramic matrix is developed by Ken-materials group of Japan Fine Ceramics Center. Micro cracks of the order of 0.01% in concrete structure are easily detected. This method can be applied to the cases to monitor damage of the structures under high temperature, in deep water or ground, or exposure to strong irradiation.

Necessity of simplification—technology suffering from spaghetti syndrome and distrusted by people
The world must aim to prosper through technology but recently there have been many doubts whether it succeeds or fails. Environmental destruction which technology has brought about and a distrust of technology itself are getting more serious now. Engineers, who are specialists, make blunders again and again as if they had no sense of technology acceptance. What the specialists developed is too complicated for people to accept. People are repelled by some. It is thought even that technology is developed only by the logic and ethics of the developers and they are self-satisfied, or engineers are insensitive to ordinary people's feeling.

There are many newly produced appliances that are hard to handle. It is stressful to use such appliances. There are a lot of examples of technology not friendly to people: one finds people washing clothes by hand because the new washing machine is too complicated to use; flooding the floor of a bathroom because of not knowing how to control the high-tech toilet. Although a household appliance or something can be used further with small repairs, repairing it is not easy in Japan today. People have a distrust of such a system and this deepens a gulf between people and technology more and more.

An accident of advanced airplanes enlarges a distrust of advanced technology. The change to automatic control of airplanes from manual control by pilots, is a problem. Since the change must be done in an emergency, complicated procedures panic the pilots. To avoid the panic at least, Yanagida thinks that the control system should be simplified as much as possible.

Yanagida diagnoses the modern technology as suffering from "spaghetti syndrome". The syndrome has five symptoms:

The *first* is a complicated appearance difficult to understand. Like spaghetti on a dish, it is hard to find where it starts, how it connects and where it ends. Ordinary people simply cannot understand the technology that is too much complicated. Though the people want to act to reduce a load on the environment, they cannot know how to do so because technology is too hard to understand. This is why they feel frustrated.

The *second* symptom is a method to apply more components to solve problems. This eventually makes technology more complicated. The *third* is one's values to consider the more complicated the more advanced. Many companies or organizations supporting R&D still omit rational technology from their targets because the technology does not seem to them to be advanced. Though a lack of wisdom causes complicated technology, they do not know it and are even proud of the complexity. To estimate degree of intelligence the speaker proposes an equation,

$$W_I = (\text{Necessary Merits})/(\text{Numbers of Components})^n$$

where n is greater than 1. We should be proud of a small number of components, the denominator, but sometimes one suffering from spaghetti syndrome boasts of large value of denominator instead.

The *fourth* symptom is paying more attention to trivial things overlooking essential matters. *Finally* technology becomes degraded by making everything more complicated. This not only is true for technology itself but also leads to distrust from people. This effect needs to be thought of as being very serious.

Techno-democracy and an NGO, Port for Techno-democracy

The strategy to make the relation between technology and the people better is to make technology as simple as possible. Also to enjoy the fruit of technology with the public, Yanagida established an NGO, Port for Techno-democracy on December 21, the winter solstice, 1996. The aim is technology of the people, by the people, for the people.

What the organization does is as follows:
1) Studying to develop rational technology
2) Selecting and praising rational technology
3) Researching and developing rational technology
4) Supporting R & D of rational technology
5) other necessary action to promote techno-democracy.

The first rational technology was selected on the winter solstice in 1998 and soil ceramics of INAX and Co., Ltd. and "Bio-coal" of Hokkaido Industrial Research Institute were praised.

Researchers or engineers should be excited by technology friendly to both people and the environment. Looking at the face of the person who is lively makes us happy. If researchers and companies act openly and happily, the people will not be alienated from technology any more. Steady communication with the people improves the relation between technology and the people. To communicate with the people, it is necessary to make technology as simple as possible. Law would not accept the technology developed without communication with the people in the near future. The technology or companies refusing communication would be boycotted. Isn't this the global standard?

Yanagida wants to gather the cases of being troubled with too complicated technology and to offer them to the developers or companies as action of the NGO, Port for Techno-democracy. He also wants to show what technology should be developed for the next millennium by praising technology for the rationality and simplicity.

Yanagida also wants to find the technology having spoiled technology, that is, the technology which developed like self-multiplying and went the wrong way, as an inside problem of technology. The organization aims to construct virtuous technology and get the system ready for it.

Inter- and Intradisciplinary Research at the Max-Planck-Institut für Metallforschung ▶▶

Manfred Rühle
Max-Planck-Institut für Metallforschung
Stuttgart, Germany

The Max Planck Society (Max-Planck-Gesellschaft = MPG) is subdivided into three sections: the Chemistry-Physics-Technology Section, Biology-Medicine Section, Arts and Humanities Section and encompasses 85 institutes which are spread all over Germany. Two institutes are located in Stuttgart: one for Metal Research and one for Solid State Research.

The headquarters of the MPG are located in Munich (General Administration and the President's Office). The President makes all decisions with respect to appointing directors and allocating funds.

The institutes are partly independent according to the new budget order. Each institute is appropriated a certain amount of money which can be spent at the institute's discretion. The amount spent for personnel, however, must be carefully controlled so that it does not exceed a certain portion of the total budget.

The contract of a director states that he/she is free to choose his/her research objective and carry out his/her research. This freedom is meant to encourage cutting-edge, pioneering research.

Due to several retirements at the Max-Planck-Institut für Metallforschung (MPI), a review committee changed the structure of the institute to a departmental structure. There will be a total of nine directors (six directors are currently at the institute). Each director heads a department and is free in carrying out his/her research objective. He/She receives a guaranteed amount of funds for his/her research. Every two years a committee reviews the activities of the institute. After a six-year period there is a major review by an expanded committee comparing the activities of different institutes working in similar areas.

The institute can also solicit outside funding from institutions such as the German Science Foundation, Federal Ministry for Education and Science, Volkswagen Foundation, etc.

The competition for soft money is rather fierce, especially between universities and Fraunhofer institutes, which are far more dependent on this source of funding than Max Planck Institutes. This often causes some strain, which can only be overcome by personal interactions.

The "headline" of the research activities of the MPI for Metal Research is

Fundamental Aspects of New Materials
The activities comprise processing, microstructural characterization, measurement of properties and theoretical aspects. The activities should concentrate on specific materials resulting in an "intra" disciplinary approach to research.

The MPI received also funding for a "Graduiertenkolleg" (i.e. a group of 10 students and 2 post-docs) where 8 University departments and 1 department of the MPI are collaborating in the area of interfaces in multi-component crystalline materials. This "Graduiertenkolleg" comprises physics, chemistry, physical chemistry, and engineering and aims to bridge the gap between science and engineering. It was quite interesting to follow the development of this "Graduiertenkolleg" since in the area of science and engineering different languages are being spoken, which makes it sometimes difficult to understand each other.

The collaboration between the students works very well. It is often more difficult to bring the professors together since they are often too busy.

The German government strongly encourages interaction with industry. In Stuttgart the MPI for Metal Research has many cooperations with Robert Bosch GmbH, as well as Daimler-Chrysler. However, the issue of intellectual property rights often negatively affects the fruitful collaboration with industry.

Bureaucracy is kept at a minimum within the MPG, much lower than at universities. Nevertheless, bureaucracy is on the rise due to many new safety and environmental protection regulations.

Currently, international Max Planck research schools are being set up which would allow an education in English, since fluent knowledge of German is often a hurdle for foreigners to become graduate students and get a PhD degree.

The MPG has a small company (Garching Innovations) which deals with patent right and marketing of technologies with industrial applications. Royalties are divided equally between the MPG, the inventors, and the institute where they work.

References
Jahrbuch der MPG 1995, pp. 467-484
Image Brochure of MPI for Metal Research
75 Years Max-Planck-Institut für Metallforschung: A Short History and Outlook for the Future.
M. Rühle, Z. Metallkunde 87 (1996) 816-826.

How Japanese Fiber-Communication Isolators Dominated the World ▶▶

Shigeyuki Kimura
National Institute for Research in Inorganic Materials
Ibaraki, Japan

In 1977, we established in NIRIM a new growth technique of single crystal YIG, yttrium iron garnet [1]. The grown crystal was meant to help improvement of microwave devices. But it turned out that the crystal was perfectly suitable for optical isolators timely requested in the optical communication technology. The first isolators in the market were produced using the YIG crystals made by the NIRIM technique. Before long, single crystal YIG was replaced by thin film crystals. But during few years of single crystal YIG domination, the market deeply accepted the basic function of the YIG isolator. The fiber communication system became to rely heavily on this device. Since this started in Japan, the world market of optical isolators has been, and still is, controlled by the Japanese manufacturers. How did this happen? The story behind is related to the research activity in NIRIM.

Before my coworkers and I ran into the YIG crystal growth, it was a common sense that crystals without congruent melting points had to be grown from solutions. Often low-melting-temperature solvents called ?flux? was utilized. YIG melts incongruently. So, it was grown from the PbO based fluxes. The flux growth was time consuming. Besides, the grown crystals were optically inhomogeneous mainly due to the impurities from the flux. Something had to be done. My background on phase diagrams, developed at PSU, made it possible [2] to establish a technique, a Floating Zone (FZ) technique, for growth of YIG crystals without a flux.

When we started the work in 1975, the target was to provide better crystals for microwave devices. Soon after our success in FZ growth of YIG, we tried to attract attentions of the industry by visiting around from one company after another. Nobody was interested. Then we asked newspaper journalists to visit us and make a report on our accomplishment. This operation was successful and a group belonging to a microwave device manufacturer came to see our result enthusiastically. Soon after, a joint operation started between this company and NIRIM. The name of this company is well known today, but it was then a small one. This company is called TR here for convenience.

The joint operation was going technologically perfect. So, we asked the technology transfer organization called JRDC, which was supported by the Japanese Government [3], an aid for the further joint operation. JRDC decided, after a hard negotiation, to help us and agreed on funding the company. But a problem arose. It was found that TR was not eligible to receive any government financial support because of a violation against an anti-communism export control rule. TR had exported to Soviet Union some industrial goods, which were in the embargo list. The company did it in a desperate effort of avoiding the bankruptcy. The effort proved insufficient and TR had to go into the subsidiary network of a Japanese computer giant. My involvement was just after the take-over. According to the government, however, the name in the black list could not be easily erased, no matter who at the moment controlled the company. Besides, the NIRIM patent could not be readily transferred to TR. The solution we found was to ask a brother subsidiary company, a ferrite manufacturer called FEC, to work together in our joint operation and to function as a channel to receive the JRDC fund.

During our effort for finding the technology transfer client, we found the possibility of using YIG single crystals for a one-way passage of light beam, an optical isolator. We quickly studied the required properties of crystal elements in such a device. We constructed a simple optical circuit for measurement of some important properties. Our background in optics was poor. So we made a tour again, this time visiting around for well-known optics laboratories in universities, government institutes and even in some telecommunication companies. We learned a lot in both, optics and the up-to-date status of telecommunication technology.

Fiber-communication people were already in business. The transmission bit-rate was relatively small then and the optical isolators did not have to rely on the magneto-optical effects. But the people were aware that a mag-neto-optical isolators would be essential before long and afraid that there would be no manufacturers of crystals with satisfactory properties even when the demand started. This situation in the market excited FEC. FEC decided to become a supplier of YIG crystals for the isolator, while TR remained interested in the crystals only for the microwave devices. The JRDC support continued for 3 years. The joint project ended in 1980 and FEC immediately started supplying high-quality YIG crystals for isolators. The supply was just in time for the high-rate optical communication. FEC enjoyed the position of the only YIG supplier for some time until YIG film crystals grown by the liquid phase epitaxy became available in the market. The YIG film for isolators was developed mainly in Japan due to the price reduction pressure from the communication market. The pressure also influenced other facets of isolator business and induced com-petitions both within Japan and between US and Japan. The further story may be given in another opportunity because of the limited space.

Through the story, one recognizes importance of communication with people in different disciplines. In many cases, barriers, no matter how formidable they may look at the beginning, can be eliminated through cooperative efforts. It is also emphasized that the interdisciplinary communication helps induce new ideas in the laboratory work.

References

[1] S. Kimura and I. Shindo, J. Crystal Growth, 41 (1977) 192

[2] Shigeyuki Kimura and Kenji Kitamura, J. Am. Ceram. Soc. 75 (1992) 1440

[3] JRDC (Research and Development Corporation of Japan) has now become a part of JST (Japan Science and Technology Corporation) and its activity can be found at the web site of http://www2.jst.go.jp/jst/develop-e.htm

Development Of Interdisciplinary Degree Programs In Materials ▶▶

Yuri Tretyakov
Department of Materials Science
Moscow State University

Public and political appreciation of the scientific community in the USA for its wartime accomplishments, particularly with respect to radar and the atomic bomb, resulted in considerable ten-years (1947–1957) support of university research mostly in terms of conventional educational structures.

The success of the Soviet Sputnik in October 1957 generated a type of shock wave in Washington decision-making authorities. Among the most important consequences was the huge–about 3 times increase of science funding and development of new university interdisciplinary materials research and materials education programs.

Traditional structure of American universities included humanity+science+engineering departments proved to be the most favorable for the development of Material Degree Programs based on the traditional engineering departments like metallurgic, mechanical, electrical, chemical engineering etc. When I had a chance to carry out my research in the USA in 1967–68 (at OSU and Penn State) and to travel around the country I saw many Materials Research Centers, Materials Research Laboratories or Departments of Materials Science and Engineering which were extremely active and corresponded to the second alternative in terms of Prof. Roy definition []. As to the former alternative, the first and wholly new Materials Science Degree Program was started in 1960 at the Penn State under compromise title "Solid State Technology".

In those and next years (say until 1986) soviet educational and research systems used to be under strong ideological pressure and were allowed extremely small flexibility, needed for dynamic multidisciplinary development. Educational system included a big number (hundreds) extremely specialized technological institutes, like Institute of Metals and Alloys, Academy of Fine Chemical Technology, Academy of Textile Materials and Processing, Institute of Electronics Materials etc. Considerable part of educational programs in these technical universities was classified and excluded openness, flexibility and international interdisciplinary cooperation. At the same time many of them were tightly connected with applied research institutes of military complex and were supported very well from the national budget. A big variety of materials educational programs developed in soviet technical universities proved to be extremely specialized and ideologized. The best part of their graduates as well as academics were oriented to military industrial activity. Another (and much smaller) part of the soviet educational system consisted of a few dozens of classical (real) universities which had a high level basic science such as physics, chemistry, mechanics, biology, mathematics, weakly developed applied science but no engineering at all. Their graduates were oriented to research institutes of the USSR Academy of Sciences, which were supplied by modern equipment much better in comparison with the classical universities but much less compared to the military oriented research institutes. It was some sort of paradox resulted in stagnation because of impossibility to realize in terms of classical university the whole set from basic to applied science and finally to engineering. It meant that basic science achievements of classic university academics had relatively weak influence on new technologies and advanced products development, in spite of the fact that all Russian Nobel Prize winners (Landau, Tamm, Semyenov and others) were university professors.

The so-called Perestroyka started in the former USSR in 1985, gave opportunities for fast development of our society, including research and education systems. It was not quite by chance that Prof. Pimental's Committee Report, published in the USA by National Academy Press in October 1985, was immediately translated in Russian and became a starting point for creation of similar Soviet Committee under Acad. V. Legasov, who had a full support of the President of USSR Academy of Sciences A.P. Alexandrov. As the result of this activity, the National soviet Research Program with the strong impact on new advanced materials and technologies development was created. Later on in 1992 the book by I. Pimental and I. Coonron "Opportunities in Chemistry Today and Tomorrow" was translated and published in Russian.

The first attempt to realize the new Materials Science Education Program inside soviet classical university was made in 1986. In addition to previous circumstances, two more reasons for such a step–Chernobyl disaster and discovery of completely new generation of superconductors (HTSC ceramics), made by Bednordz and Muller. These two absolutely different events turned out to be close to each other, because they showed effectiveness of interdisciplinary (twins) relationships of chemistry and physics as well as non-effectiveness of their independence. The Program mentioned above and called as "Advanced Materials and Technologies Program" was realized at the MSU Department of Chemistry due to initiative of Prof. Legasov. Since 1988 until now this program is being under supervision of Prof. Tretyakov. In comparison with conventional curricular of chemistry students, the original courses in mathematics, physics, solid state chemistry, materials science and materials engineering were included. This program gave a chance to graduate more than 100 students but the strict frames of departmental structure (in this case–chemistry department) didn't give a chance to realize completely interdisciplinary approach because "a combination of both disciplinary hubris and academic territoriality militated against cooperative ventures across disciplinary and administrative boundaries" (Samuel Florman; ASEE Prism 1997).

Flexibility of the whole political system existed in the USSR in the middle of Perestroyka time (1988-1990) created favorable conditions for more openness and flexibility of the educational system as well. It was a real honey-time for new educational institutions with no disciplinary subdivisions. That's why our suggestions to create perfectly new, really interdisciplinary educational institution dealing with materials science and matching as much as possible to the first stage of Prof. R. Roy's model received strong support of Education Ministry as well as Academy of Sciences. It was decided to realize this idea at the Lomonosov Moscow State University because:

- extremely high scientific potential of academics in mathematics, physics, chemistry, mechanics, computer sciences etc.
- no serious opposition from university staff because of absence of conventional engineering departments dealing with metallurgy, ceramics or polymer technologies.

At the same time Moscow State University academics were interested in advanced materials problems as their basic science activity. Joint decision of the USSR Ministry of Education and the Board of the USSR Academy of Sciences concerning Higher School of Materials Sciences (HSMS) creation, was issued on January 30, 1991. This school as interdisciplinary and independent department of MSU had to be some sort of the model of interdisciplinary in sense of:

- integration of relatively young talented academics belonging to the traditional departments but interested in creation completely new materials science education curricula (just from zero, what proved to be unique opportunity for them).
- eliminating the gap between MSU professors and Academy of Sciences Institutes researchers.

- internationalization of educational and research activity in much larger scale than it was before even at MSU which used to be relatively open for international cooperation even in soviet era. Starting HSMS activity we assumed that for real advanced material breakthrough we need to grow up a perfectly new generation of materials researchers who could be not just specialists but saying Prof. Gauckler's words, be generalists with practical experience as well as have excellent basic science background.

According to the MSU Scientific Council's decision the number of 1st year HSMS students was limited to 25 and in order to select the most capable young people-there was created an original entrance exams system consisting of two steps. The first, preliminary one gave a secondary school graduates a chance of taking examination in mathematics, chemistry and physics by correspondence. Limited numbers of winners (about 60 people) were admitted to the second, final step of written exams carried out at the University and under its control. This selection activity takes place all around the country and gives a chance to increase a competition from year to year (1996-4.5, 1997–7.0, 1998–9.5, 1999–12.9 applicants (A) per one position (P)). This year competition gave record A/P ratio among the MSU departments (A/P=3.1, 3.0 and 7.4 for physics, chemistry and mathematics departments, correspondingly). Moreover, HSMS has the highest ratio of non-Moscovites/Moscovites among MSU departments (96% in 1998).

Strong competition among HSMS applicants resulted in high level of freshmen and allowed applying the educational program of training with much more serious demands. The training itself lasts for five and half years. After first 4 years of training the HSMS students were able to obtain a B.S. degree, while additional 1,5 years of training led to M.S. degree and the Diploma verifying graduation from the Lomonosov Moscow State University. Every year consists of two semesters, every semester lasts for 21 weeks including 18 weeks of ordinary training, one week of examination and two weeks of research. Students' activity during every semester is evaluated by special rating system. The rating results are taken into account during the examination session. The curriculum includes compulsory, partly and fully selective courses.

First year (compulsory courses)
First semester
Calculus, 108h
Analytical geometry and linear algebra, 90h
Computer programming languages, 54h
General chemistry, 54h
Introduction into laboratories technique, 54h
Materials: past, present, future, 36h
English: intense course, 144h
History of world civilization, 36h
Physical training, 72h

Second semester
Calculus, 88h
Differential equations, 56h
Fundamental computer algorithms and numerical computations, 48h
Dynamic and statistical mechanics and thermodynamics, 126h
Chemistry of elements and qualitative analysis (including labs), 198h
English–supporting course, 128h
Physical training, 64h
Research work, 72h

Second year (compulsory courses)
Third semester
Calculus, 74h
Probability theory and statistics, 52h
Programming and computers, 36h
Physics of electric and magnetic phenomena, 126h

Organic chemistry including labs, 234h
English, 108h
Russian history, 36h
Physical training, 72h
Research work, 72h

Fourth semester
Theory of functions of complex variable, 54h
Equations of mathematical physics, 54h
Tensor analysis, 36h
Programming and computers, 32h
Quantum physics, 126h
Chemistry and physics of polymer materials, 60h
Analysis of substances and materials, 156h
English, 96h
History of Russian culture, 32h
Physical training, 64h
Research work, 72h

Third year (compulsory courses)
Fifth semester
Classical mechanics, 72h
Statistical physics, 126h
Chemical thermodynamics of materials, 54h
Phase equilibrium in materials forming systems, 24h
Thermodynamics of solid state reactions, 24h
Introduction in chemical kinetics, 24h
English, 54h
Philosophy, 36h
Research work, 180h

Sixth semester
Advanced calculus, 72h
Mechanics of solids, 72h

Introduction in solid state physics, 63h
Physics of semiconductors, 72h
Structural chemistry and chemistry of crystals, 56h
X-ray diffraction in materials science, 84h
Solid state spectroscopy, 32h
English, 48h
Philosophy, 48h
Research work, 180h

Fourth year (limited selective courses)
Seventh semester
Theory of determining correlation, 72h
Physics of superconductors, 27h
Physics of magnetic and dielectric materials, 54h
Physics of narrow-gap semiconductors, 27h
Chemical physics of solids, 40h
Chemical physics of dispersed solids, 40h
French, 54h
Modern economic theories, 2h
Research work, 180h

Eighth semester
Structural mechanics and mechanics of destruction, 68h
Physics of disordered mediums, 27h
Two-dimensioned structures and superstructures, 27h
Experimental methods of condensed matter, 48h
Electrochemistry, 36h
Materials diagnostics, 42h
Technology of inorganic materials, 42h
French, 48h
Management, 64h
Research work, 132h

Fifth year (limited courses)
Ninth semester
Introduction in deterministic chaos theory, 36h
Nonlinear dynamics approach in materials engineering, 36h
Radioisotope diagnostics, 36h
Computer simulation, 36h
Group theory and its application, 36h
Kinetics of solid state reactions, 36h
New generations of functional materials, 36h
Advanced inorganic chemistry, 36h
Chemistry and technology of superconductor materials, 36h
Crystals growth and thin film technology, 36h
Materials preparation laboratory, 108h
Heuristics foundations, 36h
Research work, 396h

Tenth semester
Biomechanics, 72h
Structural mechanics of polymer materials, 72h
Materials design, 72h
Noncrystalline (glass) materials, 72h
Medicine chemistry, 72h
Training in research institutes of the Russian Academy of Sciences, 360h
New laboratory experiment development, 216h
Research work, 106h

Sixth year (fully selective activity)
Eleventh semester
Research work, 720h
Tutor activity with freshmen students, 80h
MS thesis preparation and presentation to the state commission, 200h

In 1998 we started to implement the Ph.D. program in chemistry, physics and mechanics of materials for HSMS graduates. This program is available for any B.S. and M.S recipients who are able to pass successfully entrance examinations in specialty and in English.

Below is a list of the special features in the HSMS program:
- small size of the department as a dynamic and relatively inexpensive model of interdisciplinary school applicable to many Russian (and not only Russian) non-technical universities.
- an original admission procedure described before and allowed to realize advanced degree program in materials science.
- continuous and overall rating system to push students for regular hardworking
- development of student's research activity under individual ("face-to-face") supervision from their freshman years.
- integration of capable researchers from mathematics, mechanics, physics and chemistry department as well as from some research institutes of the Russian Academy of Sciences under HSMS umbrella to train and supervise students.
- participation of practically all HSMS undergraduate, graduate and Ph.D. students in different research projects as a way to support students financially.
- research activity of HSMS students gives a chance for them to receive the Soros scholarship much (15 times) more often, compared to the average level among MSU students.
- considerable (30-35%) of undergraduate students' training consists of laboratory experiments under tutors' control, with 5-6 students per one tutor.
- humanitarian courses including English, French, history, economics, philosophy, heuristics, management etc. compose a remarkable part of the program.

- every HSMS student ought to make oral or poster contribution at least twice per year at the student conference if he/she does not participate at Russian or international scientific meeting with his/her own contribution.
- due to international cooperation agreements with Fourier University in Grenoble (France), Munster University (Germany), Wisconsin University (USA) and joint research projects with foreign scientists some HSMS students have opportunity to be trained abroad for 3-6 months. A limited number of Ph.D. students have 3 years training in equal parts between Fourier University in Grenoble and Moscow State University.
- competition among students by establishing of awards system including international (e.g. E-MRS and Academia Europia prizes), national, university, certain RAS research institutes and HSMS awarding prizes (personal prizes in honor of prominent Russian scientists like Ipatyev prize).
- encouragement of all kinds of publications based on the results of students' research (20 this year graduates have 68 publications, half of them in international journals)
- lectures and invited talks by foreign professors including A. Cotton, R. Hoffman, C.N. Rao, P. Hagenmuller, A. West, A. Markworth and others.

The list of current HSMS problems includes the following ones:
1. From the very beginning of HSMS activity (1991), we were promised big governmental and industrial support. The USSR destruction with consequent industrial decline resulted in miserable HSMS budget, which was equal to USD 70,000 and USD 60,000 in 1996 and 1998, respectively. Our desperate attempts to receive any support from abroad (e.g. from ISF) occurred to be completely unsuccessful. The only exception–Soros grant (USD 14.000) saved HSMS in 1994, when it was on the verge of collapse.

2. Miserable budget spending on teacher salaries, students support, computers, reagents etc. gives no chances to complete construction of the building given to us by the university administration (it costs approximately USD 120,000 and corresponds to double annual budget expenditure). For that reason HSMS students use up to now classrooms of other departments creating a lot of difficulties.

3. Extremely low level of teacher salaries and student support results in remarkable (mainly to the USA) brain drain which we try to compensate by active involvement of the Russian Academy of Sciences researchers for students training and by new students admittance.

In spite of above mentioned problems we have a right to state that new Degree Programs in Materials Sciences at Moscow State University gave a chance:

– to contribute in selection and interdisciplinary training of the most capable young people who could become creative elite needed for Russian renaissance

– to educate a new generation of researchers for whom materials science would be a profession based on solid background of mathematics, physics, chemistry, mechanics and directed on creating original materials and technologies.

– to organize interdisciplinary research activity in terms of federal integration program dealing with new materials engineering based on nonlinear dynamics, self-organization and deterministic chaos approaches.

For 8 years of HSMS existence we had over 200 students but because of very high demands not all of them were (were or will be) able to complete MS and even BS graduate programs. MS thesis of HSMS students who were able to reach this stage (14 people in 1997 and 20 people in 1998 and 1999) received appreciation of State Commission and overwhelming majority continue their education as Ph.D. students at Moscow State University, in Russian Academy of Science research institutes and in foreign universities (mainly in the USA).

Possible evolution of MSU graduate program in Materials Science:
- expanding interdisciplinary links to include life, ecology and earth sciences (biomaterials, wastes and raw materials, earth materials etc.)
- increasing international cooperation and exchange programs with materials education institutions of European, American and Asian universities.
- converting HSMS in self-generating system allowing to bring up not only materials researchers but also materials science educators.
- development of individual training model with much more flexible education and research program.
- development of nonlinear education model starting from basic sciences and societal concern simultaneously.

List of references (Yu.D. Tretyakov)

1. "Materials Science and engineering in the U.S.", Rustum Roy, Editor, Proc. of the National Colloquy on materials. PSU Press, University Park, PA (1970)
2. R. Roy, "Pedagogical Theories and Strategies in Education for Materials Research: a hierarchical approach", Mat. Res. Soc. Symp. Proc., Vol. 66, p.23-32 (1986)
3. J. Of Mendeleev Allunion Chemical Society, 35, ? 3, p. 273-400 (1990), a special issue dealing with chemistry and materials education problems (in Russian)
4. "Ceramics and Society", R.J. Brook, Editor, Discussion of the Academy of Ceramics Forum'92, Assisi, Italy, Publ. By Techna, Faenza (1995)
5. I. Melikhov, Yu. Tretyakov. Independent Newspaper, 12th November 1996 and 7th January 1999 (in Russian)
6. Loren R. Graham "What have we learned about Science and Technology from the Russian Experience", Stanford University Press, 1999

Experience With I³R
in Materials Research:
U.S. Programs

The Interdisciplinary Labs in Materials Science...a Personal View ▶▶

Lyle H. Schwartz
Aerospace and Materials Sciences
AFOSR, Arlington, VA

Sputnik Sends a Signal

- The Manhattan Project demonstrated that the Government could respond to technical challenges
- Advanced Materials were recognized as a critical strategic need...but...
- U.S. Universities were focused on metal processing and metallurgy; stove-pipe thinking abounded; many areas of research were not on the disciplinary "approved" lists for university activity
- Industry models such as Bell Labs and GE had demonstrated what could be accomplished by teams linking strong disciplinary efforts

Multi...vs. Inter...

- Block grants to several universities were let by DARPA, encouraging local management and creativity, creating the Interdisciplinary Laboratories
- These IDL's then funded individual PI's in many departments
- At Northwestern, this meant faculty were funded in chemistry, physics, metallurgy (to be renamed materials science), chemical..., civil..., mechanical..., and electrical engineering
- This was clearly an area of multidisciplinary research on materials, with little or no encouragement of team efforts

The Seeds were Sown

- Leadership usually came from one strong department, but with different disciplines leading at the various IDL's—then strength was recruited in other departments
- Community and cross-departmental interaction was stimulated through:
  - collective management of resources
  - central sample preparation/characterization facilities
  - discussions of research progress to build common language and philosophy

...and Began to Grow

- **Materials** was identified as a national need and a challenging and fruitful area for academic research
- Other Federal agencies expanded their programs in all aspects of materials and a field of endeavor evolved
- Other universities sought ways to emulate the IDL's-if not through block funding, through organization and stimulation in hiring and collective management of resources
- Educational programs were initiated or transformed to reflect the newly emerging intellectual field

Multi...became Inter...

- The Mansfield Amendment moved DOD out of "civilian" research in 1972
- The IDL's were transferred to NSF...but how could they be differentiated from the conventional NSF single investigator grants?
- Roman Wasilewski and the concept of Thrust Group came to the newly named Materials Research Centers
- Most of NSF and most faculty fought the new concept...after all, *everyone* knew that "seminal" ideas came from individuals, not groups

...and Thrived

- Slowly, cautiously, cross disciplinary groups were formed to attack complex problem areas
- At Northwestern, an additional strategy was developed to counter concerns about tenure and promotion-joint appointment in multiple departments
- Interdisciplinary research spread beyond the MRC block grants *via* other-agency funding, and attracted the best scientists because their work prospered and was enriched by continual stimulation at the interfaces between the disciplines

Everybody Liked the Good Idea...Eventually

- It took ten years (and Erich Bloch) before NSF understood what a wonder they had wrought
- Now the model of the MRC's has propagated to offspring such as the Science and Technology Centers, the Engineering Research Centers,...etc.
- *Materials Science* became *Materials Science and Engineering* at many universities...and more than one hundred related departments with varied names now exist in the U.S.
- The field of endeavor was invented in the U.S., but is now significant the world around
- **Materials** is almost always listed-along with biology and information-as a critical technology for the next century

And So to Sum Up...

- Materials research at universities, encouraged by an enlightened few in Federal funding agencies, evolved into an interdisciplinary effort in which rich and exciting technical advances continue to be made
- Time passed and the field of materials research penetrated deeply into the more traditional "disciplines"

- Enough time has passed so that the academic departments which grew in the midst of this research endeavor have now themselves created their own "discipline"...but are still wrestling with its elements and boundaries

Links to Industry Began...
- We began with the question posed by many faculty: What can industry do for us?...meaning ...how can we get some $ from them
- Individual faculty consulted, some extensively, and brought industrial problems into the academic sphere for examination...but...
- Industrial problems rarely are disciplinary, so it was not until groups of faculty began to work together that industry could see universities as a source of technical input as well as the farm teams for future players in the industrial big leagues

...and Grew
- Centers sought industrial advisory committees
- These committees became the conduit of information between the university research and the industrial need
- As many industries shed their own long-range research in materials, two complementary trends developed:
  - Researchers leaving those industries in many instances migrated to universities
  - Industries sought out these same individuals who now address many of the industrial needs for long term research while located in academic centers

Is it all so Rosy?
- University research cannot substitute for industrial efforts
- Transfer from the company laboratory to the manufacturing scene was hard enough in the "good old days" and its harder yet when the gap between university and industry is added

- As industry sheds its own research knowledgeable people, the ability to develop a community of interest which includes university and academic scientists becomes more difficult…exaggerating the gap

The Big Challenges Remain
- Clarifying the difference between materials research-the research endeavor, and materials science and engineering-the "discipline"
- Bridging the gaps between academic product and industrial need

…and There's One New Concern
- Interdisciplinary research was initiated by DOD
- Although the IDL's moved to NSF, DOD remained a major funder of significant elements of the field
- And now with drastic cuts in DOD S&T funding, the health of those elements is threatened

Comments on Interdisciplinarity—Past and Future ▶▶

Paul C. Maxwell
Vice President for Research and Sponsored Projects
University of Texas at El Paso

Driving Force(s) for Change

- Question: What are the origins of MSE?
  - Gang of four at Penn State
  - Others at MIT, Stanford, etc.
  - DARPA
- Answer: Yes, but...This was a national change brought about by a major event.

Sputnik*

- Broke our technological complacency
- Awakened fears of an uncertain future in the "ideal" world of the 50's.
- Lead to political claims of a "missile gap"
- NSF funding for education accelerated—"new math invented".
- JFK announced the "race to the moon".
- Need for new materials recognized by Fred Seitz, others; Congress and the Administration agreed.
- Materials Science & Engineering was born!

* "Probably no other single event since then has had as much influence in causing a reorientation of thinking (and action) concerning federal

support of research and education"—Julius J. Harwood, Materials Science &Engineering in the United States, ed. by Rustum Roy, 1969.

Fast Forward to 1999:
* We won the space "race".
* The Cold War ended—no more bad guy (at least at first) to drive research.
* New World of Global issues.
  - Global Economies
  - Global Climate Change
  - Global Communication
  - Global Health Concerns (AIDS, TB, Malaria, etc.)
  - Global Terrorism/Crime

Need to Change "Triangular" View of IDL to "Tetragonal"

Ongoing IDL Activities at UTEP
* Materials Corridor Initiative
  - ~$1.8 Million with Mexico
* Internet 2
  - ~$2.5 Million in State funding
* Border Health Initiative/Institute (BHI)
  - $50 Million in State funding

Materials Corridor Partnership Initiative
Sustainable Economic Growth in the U.S.-Mexico Border Region

A Proposal for Action-The May, 1997 BNC
* During the May 1997 Bi-National Commission meeting, Mexico called for creating a "Materials Corridor" Initiative.
* The Initiative would be a full research partnership between our two countries.

- Envisioned creating a materials "Silicon Valley" along the U.S.-Mexico Border.

Relevant Background

- *2,000 mile border* is center of manufacturing and materials industries crucial to both countries.
- In El Paso-Juarez alone there are 415 *maquiladoras* industries.
- ~80% of total industry in the State of Chihuahua is materials related.
- Mexican Materials industry ~ $6.8 Billion along the border.
- Major materials academic and research institutions in the border region, including:
    - UTEP, UNM, NM Tech, UCSD, UC Riverside, ASU, etc. (U.S. side)
    - UNAM, Monterrey Tech., IPN, etc. (Mexico)
    - Los Alamos, Sandia, Berkeley, etc. (U.S.)
    - CIMAV, CINVESTAV, COMIMSA,, ININ, etc. (Mexico).
- Major U.S. Investment in Materials R&D
    - ~*$2 Billion* annually in U.S.
    - Major programs in *DOE, NSF, NIST, NASA, DOD, etc.*
- Major Border environmental programs:
    - EPA Water infrastructure *($740 m)*
    - CNA Water treatment, etc. *(~$150 m /yr.)*
    - BECC/NAD Bank *(~$3 Billion)*
    - IBWC

Materials Corridor Initiative: Sustainable Development to Improve Energy Efficiency and Reduce Emissions

- Partnership between Research/Academia, Industry, and Governments.
- Focus on energy efficient, with clean, sustainable materials industries on both sides of the border.

- Potential for major impact on energy use, emissions and pollution control.
- Basic to advanced materials (Petro/Chem., Elect., Manufacturing, etc.).
- Human Resource Development.

Private Sector Participation—BCSD
- Business Council for Sustainable Development—Bechtel, Conoco, Chaparal, Dupont, Grupo IMSA, Pemex, etc.
- *Eco-efficiency is good for the bottom line.*
- Model program beginning in Tamulipas (Petro-chem.); possible programs in Monterrey, Juarez and Tijuana.
- Others: *Maquiladoras* industries.

Sustainable Industry Example: Chaparal Steel and TXI, Cement
- Steel and Cement Manufacturers co-located next to each other
- Waste slag was introduced as raw input to the cement production.
- Conserved natural resources reduced energy use 15% and eliminated waste slag.
- Production capacity increased 9%; Increasing profits $3 million in one year.

Materials Research Focus (tentative)
- Energy Efficient and Environmental Materials.
- Materials Processing, Synthesis and Fabrication.
- Nanomaterials and catalysts.
- Thin Films.
- Corrosion.
- Biomedical Materials.
- Synthetic Materials.
- Photonics/Electro-Optic Materials.

Materials Education
- Universities to bear primary responsibility.
- Summer internships and thesis work via labs, industry.
- Summer employment exchange program.
- Scholarships/Fellowships—undergraduate, graduate, post-doctoral.

Interested Parties to Date
- OSTP, NSF, NIST, DOE (Germantown), DOD, EPA, Conacyt, Semarnap.
- UTEP, UT Austin, Univ. Calif., UNAM.
- Los Alamos, Sandia, ININ, CIMAV, others.
- Business Council on Sustainable Development (BCSD).
- U.S.-Mexico Science Foundation.

Working Meeting at UTEP
April 16-18, 1998
- Formation of Materials Corridor Council
- Commitment by Mexico of $1 Million to initiate program.
- Preliminary proposals identified; need for additional regional work with private sector and institutions.

June 1998 Bi-National Meeting, Washington, DC
- Mexican delegation head, Carlos Bazdresch, Director of CONACYT, re-iterates commitment to Materials Corridor Initiative
  - Requires match with U.S.
  - Requires strong participation by private sector.
- U.S. accepts challenge to find funds and support.

Examples of Possible Projects under the Materials Corridor
- "Green Cement"—use supercritical, liquid CO_2 mixed with fly ash to develop new light weight cement. (Los Alamos National Lab).

- "Nanocatalysts"—develop new catalytic materials to remove SO2 and NOX from petroleum crudes to meet new standards in US & Mexico (UTEP).
- "By-Product Synergy"—work with BCSD to develop 4-5 projects in border area to involve ~12 manufacturers per project focused on waste utilization and eco-efficiency. Private sector to match ~$100K per project (BCSD).
- Advanced Colloidal Suspensions: New technology developed originally to remove heavy metals from electrolytic solvents in electroplating, for use as new municipal water treatment technology (Los Alamos National Laboratory).
- Vanadium Recovery: Land fill removal for use in Steel Industry (UA).

"Green" Cement

- Use supercritical CO2 and fly ash to create new, super "Green" cements.
- Technology takes uses waste heat in combination with other waste byproducts of coal-fired power plants to produce new, superior cement.
- Advantages:
  - Permanent sequestration of the greenhouse gas, CO_2.
  - New cement materials cycle eliminates creation of CO_2 (current cement production produces X% of industrial CO_2).
  - Variation of ash content can change vary mechanical/physical properties of cement from 2X-4X normal strength to 1/4 normal weight at regular strength.
  - New cement is impervious to water penetration.
  - Creates new sustainable markets and economies.

Nanocatalysts

- Exxon, Shell, Pemex are important border energy industries.
- Conversion & upgrading of increasingly heavy petroleum feedstocks is required.

- Better removal of sulfur from transportation, power and heating fuels needed. Also applicable to synthetic fuels.
- Science of transition metal sulfide nanocatalysts is well understood, but—
- Additional materials research needed for commercialization.
- DOE synchrotoron facilities important to success already have been used in initial experiments.

Congress Supports Initiative

- Senator Jeff Bingaman (D-NM) and Congressman George E. Brown, Jr (D-CA) introduce the "Materials Corridor Partnership Act of 1999" (HR 666 and S.397; Feb. 10, 1999)
- Companion bills call for:
  - $ 5 million per year for 5 years
  - Funding to DOE to be distributed among appropriate agencies, including NSF, NIST, EPA and State.
  - Focus on "energy efficient, environmentally sound economic development along the U.S.-Mexico border through research, development, and use of new materials technology."
  - Creates special advisory committee to the Secretary for use of the funding.
  - Strong involvement of US and Mexican private sector emphasized.

"Continued progress and innovation in materials research lies at the heart of our scientific and technological future."

OSTP Report to Congress, "Science and Technology Shaping the 21st Century," April 1997.

Personal Observations on Interdisciplinarity ▶▶

E. N. Kaufmann
Argonne National Laboratory

Introduction

This somewhat extended abstract is being composed after having benefited from attending the I³R Meeting and having enjoyed the wide variety of perspectives and experiences communicated by so many knowledgeable participants. You might think that would make this an easier task, but the opposite is true. My "assignment" was to report on how I³ relates to professional societies and journals I have known, a not unreasonable request given past associations with the MRS and its journal, JMR, particularly in their more formative years. Some recollections and some comments on current postures of MRS and JMR will be found in the section following. To be sure, there are manifold anecdotes one might relate about overcoming (or not) barriers raised by disciplinary preconception and much revered institutional norms. But to what end? On recalling my own involvement and on trying to discern the common elements in all I heard at our Meeting, I conclude that lessons learned from such accounts are, at the detail level, too situation-specific to be generally useful while at the same time being easily generalized to a few tenets that most of us by now find obvious in principle but that provide no actionable roadmap for implementing I³ in a specific new arena. How can that be?

Other contributors are submitting the I³R challenge to scholarly analysis and reporting on significant impediments and enviable achievements. In such a context, my observations at best may seem overly simplistic and could easily come across as a glib trivialization of the whole endeavor. Nevertheless, I cannot resist noting that the common themes permeating our entire discussion reduce to a few fundamental aspects of human nature well known to us all and ubiquitous not merely in our universe of science, technology and research, but in broader society. I feel it is important not to lose sight of this as one examines I^3 problems and solutions, for it is often the larger context that rises to thwart the best of local intentions.

Therefore, let's say what goes without saying. We are a risk adverse species. This translates into resistance to change and thus to institutional inertia. When taken in concert with our subjective propensity to categorize and the objective need at any given stage of development to parse complex systems into manageable sub-units, it becomes clear why rigid taxonomies are the rule. Be they definitions of departments on campus, political labels, or funding program classifications.

It is fair to say that in the examples presented at this gathering, the origin of hurdles confronting introduction of an I^3 approach in extant systems is the inevitable rigidity of existing structures. Similarly, successes seem to arise when one or more of a few criteria are met. Based on enlightened self-interest, the principal movers find the risk-reward calculus of success compelling enough to pay the price of breaking with tradition. This is often facilitated when the entrepreneurial venture does not risk the mainstay of the principals' vocations. Also, barriers are lowered considerably when the new I^3 enterprise is "green field," i.e., not making a frontal assault on a preexisting structure. Otherwise, the organizations involved must offer some open avenues to change or at least to circumvent boundaries. Having the right-sized, well-positioned resources is certainly also necessary.

Of course, it would be naïve to underestimate the importance of the details underlying any and all I^3 successes, but these must be devised in the milieu of the particular people and institutions involved. I believe no generic protocols for barrier reduction or for incentives that balance risk will fit differing circumstances. Attempts to map one successful formula onto a new situation are likely to disappoint. Likewise, general programs to promote I³ in a top down fashion may raise awareness and enthusiasm, but are always tested at the bench—just as in technology transfer, I³R is a contact sport. Although not sufficient, it is nevertheless still clearly a necessary prerequisite to create and sustain genuine and practical inducements that mitigate the risk attendant to disturbing the status quo and blunt institutional disincentives.

MRS and JMR

There is no doubt that the Materials Research Society at its founding and to the present succeeds in crossing disciplinary boundaries with impunity. Success is defined here as producing programming across the science and engineering disciplines that is very well received by "customers" from all these disciplines. The last phrase is emphasized because attempts to mimic the MRS formula over the years have done less well when measured against the interdisciplinarity of the audience as opposed to that of the program per se.

How do the above generalizations apply to the nucleation and growth of MRS? The recognition that materials research in the real world was by its nature interdisciplinary (a premise already accepted in practice in industrial labs and in principle in some forward thinking enclaves in (D)ARPA and NSF) and that existing society forums were not adequate moved the MRS founders to offer a forum focused on topic regardless of discipline. The reward would simply be a more conducive forum for their own, their students', and their colleagues' research reports. The risk (i.e., penalty for failure) of this ancillary activity was virtually nonexistent and the required resources minimal. The MRS affinity group did not forsake their personal single discipline societies, so very little territorial backpressure on growth from traditional organizations was engendered. This is an excellent example of how, in the absence of personal and institutional barriers, a good idea with devoted backers will prosper and grow to a size sufficient to then survive being noticed by more traditional players.

MRS has so far successfully resisted attempts to "divisionalize" its programming and continues to rely heavily for its I^3 personality on a healthy high-turnover-cadre of enthusiastic (low-risk/high-reward) volunteers. Many new generation researchers consider the materials research field and the MRS, respectively, as their primary discipline and society. After a quarter century, today's challenge for MRS is continued vigilance against the threat of the creeping rigidity that often accompanies an organization's transition from an entrepreneurial to a bureaucratic stage.

The Journal of Materials Research (JMR) is a somewhat different story. MRS launched JMR in January 1986 after performing for several years all the due diligence studies one would normally expect for a new publication. Suffice it to note that the mechanics of business plan preparation were straightforward. These were pursued while MRS was still in a highly innovative entrepreneurial phase, actually coinciding with the establishment of its first stand-alone headquarters office and modest full-time staff. One admittedly self-serving rationale for the young MRS to launch a journal was the idea that a society is not really a society until it has its own journal. Materials research at the time was being touted in many quarters as the emerging field, a real growth area. More mature discipline-oriented organizations were therefore anxious to collaborate (i.e., collegially joint venture) with MRS. The American Institute of Physics was thus involved from the outset as JMR's first publisher. So design, production, marketing, fulfillment, i.e., all the practical production considerations met no more challenges than any new publication might expect. There were few, if any, internal or external resisters of change to contend with. Where then did the I^3R character of JMR enter the picture?

Those still influenced by a single-discipline mindset saw the great success of MRS meetings and their rapid growth and from these somewhat superficially concluded that its journal would automatically enjoy the same success. Of course a meeting and a journal are different things. MRS's own community survey hinted strongly that presenters at MRS meetings would first choose to publish a finished archival research paper in their own favorite single-topic or single-discipline journal. And, typical MRS meeting attendees would first invest their valuable time reading, those same, more narrowly focused journals. The dominant reasons, i.e., that authors want to target their specific audience and that readers want to efficiently find and digest new developments within their own limited interests, were not the only reasons. The imprimatur imputed to the chosen periodical anoints its pages with a credibility and apparent significance that no new upstart journal could offer.

MRS was looking at several years of red ink that might never be recouped, but to its credit and after much soul searching, the ideal that JMR should reflect the panorama of materials research, even well beyond that common at MRS meetings, prevailed over the quite legitimate fears of the "focus or fail" proponents. JMR is now a fiscally sound, widely cited, well-respected journal, able to maintain high standards of paper acceptance. True, some partial focusing strategies along the way to "capture" topics under-represented elsewhere have been employed. It is also true that, whereas JMR has become the primary outlet for some, it will likely never be that for the majority of narrow topics and disciplines that fit under the materials field umbrella. I regard the initial decision to remain broad and MRS's continued comfort with JMR striving to be the best reflection of the field, even while being number two for narrower purposes, to be a victory for the I³R ethos.

A Conclusion of Sorts

Perhaps it is clear that I am not a fan of specially propounded programs to affect I^3R. These I feel are all too often only of two types: (1) ineffective because words are not supported by real resources and other necessary organizational follow-through, or (2) over-defined such that these programs themselves evolve along their own rigid lines allowing at most one generation of real interdisciplinarity. Common underlying aspects of I^3 successes always seem to include a bit of chaos, lack of definition, and an unfettered settling-in process for collaborators that allows only the favorable and productive connections to form "organically." This is tantamount to suggesting that non-structured formats, deconstructed taxonomies, multiply degenerate ground states, and continually moving research targets should be the rule.

A small dose of realism, of course, tells us that this highly creative and disorganized process must somehow couple to the larger very structured world of the ultimate customer. Therein, I suggest, lies the key clue to ultimate success. I believe pockets of I^3R "Utopian" anarchy can actually be sustained, even within and between our structured institutions when enlightened intermediation is present. It's clear from the presentations at this gathering that many here have played just this critical role in their own stories–that is they manage the interface so as simultaneously to protect innovation from undue premature interference, to provide an overall sense of direction without directives, and to satisfy the sources of resources that their long term goals are being served. A very tall order and probably the reason that we here hope to discover some best practices at meetings such as this.

Appendix–Leftovers

A plethora of disjoint issues concerning I^3R and its context remains on my figurative notepad. These could be characterized as assorted symptoms for which I don't pretend to have the cure nor even always have a clear idea of the root causes. Some were topics of at least brief discussion at our Meeting while others were overlooked. They are enumerated below as mercifully brief, albeit occasionally irreverent, snippets simply to encourage their further consideration in other future forums..

The "applied" versus the "basic" character of R&D within and between institutions are often are made rivals rather than partners. Basic folks resist being mere job shops while applied folks resist funding sand-box or blue-sky research.

As an editor, I decry the rapid disappearance of industrial researchers whose managers fail to reward, or even tolerate, time spent on scholarly works.

Ubiquitous misunderstanding of the multiprogram DOE national laboratories (conspicuously under-represented at our Meeting, as it happens) leads to their being simultaneously praised for addressing large complex multidisciplinary projects and chided by blue ribbon committees for not having been more narrowly focused on missions so that they might shed some of the very breadth on which their utility rests.

Workers in disparate fields with very different institutional goals speak different languages. Jargon is readily overcome but drastic differences in unspoken context will scuttle a collaboration if no one notices that participants are "talking past each other."

Equitable assignment of the rewards of I³R can be another serious impediment.

How personal credit is shared within teams and attributed to individuals in their home institutions will determine how collaborators interact.

Intellectual property rights, a more concrete reward allocation issue, can scuttle collaborations when legal conservatism controls what ought to be primarily business decisions. Included here are the fuzzy notions of "pre-competitive" and "generic" research that supposedly avoids IP concerns. Does it then also avoid valuable commercializable discovery?

Interdisciplinarity requires some breadth in the topical expertise of each collaborator, but as with any investigation, forefront research demands deep understanding of the fundamentals and the details of each component task. How does one balance these apparently contradictory traits as reflected in curricula and post graduate training? Put another way, can a "renaissance researcher" fill both bills?

Materials research tends to be owned or disowned from time to time by the disciplines, depending on whether credit for a materials discovery may be claimed. Typically, explicit acknowledgement of sister disciplines' role is conveniently overlooked. Is there a substantive downside for the image of the field in the eyes of both students and funding sources because this otherwise somewhat amusing game?

Do slogans and buzzwords such as "coopetition" or "colloboratory" help raise awareness and communicate the I^3 idea more accurately and succinctly, or are they mere advertising gimmicks?

Can anyone claim a net efficiency in I^3R over less collaborative algorithms? I.e., is the extra cost in time and effort to needed to overcome traditional resistance to crossing those boundaries, outweighed quantitatively by increased value of results and/or increased productivity of the team?

Even accepting the practical barriers to actually accomplishing an I^3-type arrangement, why is it that we must continually re-explain why the concept per se is good? There are many things in life that are difficult to do but are obviously good for us if done. Why is I^3 different?

Experience With I³R in Materials Research: Local

Barriers to Organized Interdisciplinary Research in a University Environment ▶▶

Forrest J. Remick
The Pennsylvania State University

I write from the perspective of one of the few individuals remaining who was involved in some of the earlier days of the administration of organized interdisciplinary research programs, or what has become known at Penn State as Intercollege Research Programs (IRPs). In fact, I served in this capacity for twenty-two years from 1967 through 1989 until appointed by President George Bush as Commissioner of the US Nuclear Regulatory Commission.

It is my pleasure to participate in this Conference, but I must admit that it is a little difficult to regain fully my academic perspective, having been gone from the University for nearly ten years.

For the record, let it be clear that I am not a material scientist. However, other than for a short period at the former Bell Telephone Research Laboratories, I have been engaged for almost all of my professional career in interdisciplinary activities of one sort or another.

I can assure you that being NRC Commissioner involves one in a morass of interdisciplinary activity of science, engineering, law, public policy, politics, national security and international activity of a wide variety.

My interdisciplinary activities began when in 1956 I came to Penn State from the Oak Ridge National Laboratory to teach in the International School of Nuclear Science and Engineering (ISNSE) and to operate and conduct research at the Penn State Research Reactor, which incidentally, although receiving the second license for a nuclear reactor, is the first licensed reactor to operate in the United States and is the longest operating licensed reactor in the country. The International School was sponsored by the US State Department in an effort to make the recently declassified information of nuclear science and engineering available to scientists and engineers from many parts of the world. This was before Penn State had a Nuclear Engineering Department, but did have an interdisciplinary graduate degree program in nuclear science, which served as the basis for the ISNSE.

The nuclear research reactor was established as a University-wide research facility and has always been used by faculty and graduate students from numerous disciplines of the University as well as by industry and other universities

As an early director of this University facility, I learned the joys of and some of the frustrations and barriers to interdisciplinary activities within a university in which the traditional structure and rewards are based on a fairly inflexible and largely Balkanized disciplinary structure.

After spending nine years at the University in charge of the campus reactor, as well as the University's second nuclear facility fifty miles north of here, I took leave and accepted a position with the International Atomic Energy Agency (IAEA) in Vienna, Austria as Chief of the Training Section of the Department of Technical Assistance. In this position, I was responsible for the international nuclear science and engineering education and training activities of this United Nations associated agency.

The interdisciplinary education and training activities were performed in many parts of the world and I found that international interdisciplinary cooperation could work quite well. Scientists from various disciplines and countries worked together very well and students were largely enthusiastic about working with others of a variety of disciplines and nationalities, something I've observed many times here at MRL. However, it was when some administrators or politicians became involved that barriers sometimes where erected, not unlike what can happen within a university.

Although I have participated in interdisciplinary instruction and I am a strong believer and supporter of interdisciplinary instruction, at least at the graduate degree level, my intent is to discuss primarily interdisciplinary research, and in particular organized interdisciplinary research units.

When I returned from Vienna in 1967, I was invited by the Vice President for Research to join his office as Assistant Vice President and Director of the Institute for Science and Engineering (ISE). Dr. E.F. Osborn, better known to all as Ozzie, was VP for Research.

He was an outstanding scientist and highly respected and well liked Professor of Geochemistry and former Dean of the then College of Mineral Industries, now the College of Earth & Mineral Science. Ozzie had the complete confidence of the President at that time, Dr. Eric Walker. Incidentally, Dr. Walker was responsible for the building a research reactor at Penn State

They shared the vision for improving the stature of the University, which included rallying the strengths of the disciplines into greater collaborative teams to attack larger problems of society which could not be undertaken effectively by individual or single discipline researchers, which was largely the mode of operation of the University's research at the time.

This was the period after the wake-up call of Sputnik, and they were anxious to get the University out of its comfortable doldrums and hopefully become a major research institution

Ozzie was an informal, extremely efficient research administrator who was very effective at getting faculty to work together for the good of the University. With the ear and confidence of the President, Ozzie was able to make things happen quickly.

I am delighted to see that the conference room at the Material Research Laboratory was dedicated to E.F. Osborn's lasting memory as I consider him to be the epitome of an effective University scholar, administrator and statesman with whom it was a pleasure and honor to work.

One of their first efforts towards increased interdisciplinary research activities was the formation in 1962 of the Institute for Science and Engineering (ISE) within the Office of the Vice President for Research. Within the ISE they located the Center for Air Environment Studies, the Animal Behavior Laboratory, the Computation Center, the Human Performance Laboratory, and the Land and Water Resources Institute. They also moved into the ISE the Ordnance Research Laboratory, now known as the Applied Research Laboratory, from the College of Engineering and the Material Research Laboratory (MRL) from the then College of Mineral Industries. The Director of the MRL at that time, and in fact its charter director, was none other than our distinguished colleague, Dr. Rustum Roy

As the number of interdisciplinary and intercollege research institutes and centers were formed the ISE eventually became a subset of the larger Intercollege Research Programs, (IRP) which at one time consisted of as many as eighteen separate units representing about one-third of the University's sponsored research activities. Unfortunately, I note that this highly successful initiative has been allowed to dwindle to seven units today.

As a general matter, the directors of the IRP units are faculty tenured in a department of a college of the University. Typically, the units involve additional faculty members with joint appointments and tenure within various University departments, offer no degrees, and frequently employ non-tenured research faculty, as well as technical, clerical and administrative personnel. In fact, one of the strengths of such units is the existence of dedicated support staff to enable quick response to funding opportunities as well as provide technical and administrative support to the research projects.

IRP directors have responsibilities much like department heads and the IRP Director has responsibilities much like a dean, except they do not have responsibility for degree granting instructional programs unless this is an independent responsibility within a department or within the Graduate School.

Although some universities had individual interdisciplinary research laboratories, which frequently were located off their main campuses, Penn State was one of the first, if not the first, university in establishing a formal administrative structure and related policies for administrating a number of organized interdisciplinary research units outside the normal department/college structure. The novelness of such a structure was obvious, based on the number of universities who visited or inquired about the structure and the policies established for its operation and control.

Certainly, not all within the University favored such an organizational structure. Many lessons were learned, both good and some not so good, as a result of undertaking such an interdisciplinary organizational structure. Admittedly, some mistakes were made early on. A number of these were corrected or made right over time. However, although corrected, some have never been forgotten. And long standing animosities lingered, at least until individuals involved reached retirement or greater finality.

There is no question that the Interdisciplinary research units are an important part of the University, its research activity, and its stature and reputation and have been for the last thirty-seven years

For example, total research expenditures in these units in fiscal year 1997–1998 amounted to approximately $120 million ($119,776,000) representing approximately one-third (32%) of total University sponsored research expenditures ($374,145,000). They involve well in excess of a thousand individuals, not including the hundreds and hundreds of graduate and undergraduate students supported while working on complex interdisciplinary research. To me, one third of the University's total sponsored research expenditures and the resulting support of faculty and a multitude of students speak to the importance of these units

Yet, in spite of this I believe that the University has neither fully faced up to the importance of these units nor has made some of the administrative adjustments necessary to fully reach the potential of the units; thus, removing some of the lingering administrative barriers to their greater acceptance and opportunity.

Why do I believe this? What are the barriers to organized interdisciplinary research?

It largely boils down to barriers, which are allowed to exist as a result of the traditional organization of most large universities and the methods of selecting, evaluating and rewarding university administrators.

The traditional university organization consisting of departments within colleges reflects the time honored, although admittedly somewhat arbitrary, division of scholarly activity into recognized fields of study or disciplines.

This separation is important, for it is by close association and constant communication among scholars involved in basic studies and in relatively narrow areas that much new knowledge is gained. Even in the areas, which concentrate on the application of knowledge, specialization is important if mastery of skills and meaningful results are to be achieved.

In principle, This is all well and good inasmuch as I firmly believe that one cannot have strong interdisciplinary research programs without a strong disciplinary base to draw upon for faculty and students interested in working with those of other disciplines on complex problems of society. However, this organizational structure, as normally administered, does not encourage, evaluate or reward academic administrators for encouraging faculty and students to become engaged in academic activities outside the disciplinary bounds of a department or college, in cases where it might be appropriate and where such involvement would benefit the individual, the University and society. In fact, there are cases where faculties have suffered from salary, promotion and tenure decisions because of becoming involved in interdisciplinary activities

On the other hand, many of society's problems do not fit neatly into the University's departmental grid, nor are they rapidly divisible into sub-problems, which yield to isolated investigation and independent solution. Many require the attention of scholars who not only have the depth of knowledge and competence in specific disciplines but who also are in, and able, to communicate in related areas.

Historically, fertile areas for investigation frequently have been found at the interface of two established disciplines and have led to the growth of new ones; for example geophysics, bio-mechanics, bioengineering, etc.

There is unquestionable evidence that scholars and their students from diverse disciplines can work together effectively on common complex problems with tangible benefits to all, if careful thought is given as to how to encourage and sustain such interaction over a period of time.

Interdisciplinary or intercollege research units enable faculty and their students to readily form interdisciplinary research teams which can readily respond to multi-discipline, problem-oriented research and public service opportunities. Pulling together interdisciplinary teams, developing proposals, obtaining the multitude of clearances, etc. Is no small task, but is one that the intercollege units generally do quite well. Further, they provide visibility for the University's commitment to a particular area of research activity, such as materials, transportation, or environment, which otherwise might remain diffuse and unrecognized within the traditional University structure. Their existence, provides continuity of interest and focus to interdisciplinary research need and opportunities. Evidence of such continuity of interest, effort and track record can be extremely important to external funding organizations.

So what is my message today?

I realize that some of what I have to say may be controversial and anger some. But, I spent essentially all of my professional and University life involved in interdisciplinary activities of one type or another, including twenty-two years of administering parts or all of the Intercollege Research Programs. Thus, my interdisciplinary beard is gray.

I'm sure that you've heard it said that you could tell your old, when you have all the answers but nobody asks you the questions.

However, I have been asked questions on The Driving Forces for and Barriers to Interactive Research and would be negligent to not share them with you, as I see them at this moment, leavened by ten years largely removed from the University fray. I promise you that my views are heart felt and held with what I consider to be my interest for the betterment of this fine institution.

In short, just as strength of the disciplines is an extremely important aspect of scholarly activity, strong, complementary, interdisciplinary activities can add considerable breadth of scholarship to a university and be of considerable benefit in preparing students for the interdisciplinary environment within which most will work once they graduate. Understanding, willingness, and ability to interact effectively with those of other disciplines enhance one's chances of succeeding in the real world. How any times I've said that the University is umpteen acres surrounded by reality.

My involvement in seeing, and hopefully helping, the Intercollege Research Programs develop was a particularly satisfying experience for me. To see the enthusiasm of students as they worked together and shared knowledge with one another across disciplinary lines was exciting and rewarding. To work with the highly entrepreneurial and dynamic unit directors, such as Rustum and many others, was a pleasure and a learning experience.

As many of you know, Rustum can rattle the cages and ruffle a few feathers with his unlimited energy, impatience with bureaucracy, creativity and commendable openness in sharing thoughts and ideas. I have openly expressed the desire for and need for dozens more like him at the University. He, as others like him, make a university what it should be; a place where ideas are openly expressed and discussed and where complacency does not set in.

Well, what do I see as the needed requisites for success of organized, interdisciplinary research units in the face of the traditional discipline-oriented organizational structure of universities, which frequently views such units as a threat to the power structure of a university and a threat to the allocation of resources.

Unquestionably, success is strongly dependent on a strong advocate or advocates at the top academic levels of the university who are believers and provide leadership for both disciplinary and interdisciplinary research. Obvious disinterest or visible opposition by such individuals would constitute the biggest barrier to intercollege research.

Advocates and leaders, like those of Drs. Osborn and Walker, are essential to overcome the reluctance of those who see such initiatives as a threat to the status quo. Strong advocates are needed who will make sure that all academic administrators, disciplinary or interdisciplinary, are evaluated and rewarded in part on their cooperation and support of research across, as well as within, disciplinary boundaries. In other words, academic administrators should be evaluated on their specific breadth of University vision and interest.

Digressing for the moment from interdisciplinary or intercollege research, I wish to comment on what I consider a troubling practice of universities. One that I've seen cause adverse effects. Universities follow the practice of limiting their search for academic administrators strictly from those who have come up through the ranks.

Heaven forbid, that those not cloned from that mold should be considered as worthy of consideration.

They then place these cloned individuals into administrative positions with little, if any, preparation in such matters as supervisory skills, communication, team building, leadership, managing change, personnel matters, employee evaluation, administrative systems and processes, etc.

We academics, and I repeat we academics, scoff at any thought that there might be need for such preparation. After all, we are so smart, so capable, and so important to society that we do not need to know of such nonsense.

The view is that if a good researcher, obviously a good administrator. However, many good researchers admirably want to continue their research. With few exceptions, this inevitably results in their relegating their administrative responsibilities to the background, frequently with detrimental effects, both in efficiency and esprit de corps, within the academic unit, which they are to administer. This often results in a loss of faculty and staff allegiance to the department, college and university.

Unfortunately, faced with such administrative mistakes, universities are very reluctant and extremely slow to correct the mistake for the betterment of the academic unit and the university. It is hard to fathom why universities disregard what others have found to be so important.

Now back to Intercollege research.

The best interest of the student, the faculty, and the University should prevail over parochial concern over who gets credit for student and faculty research. At the same time, I learned early on that concern over who gets credit for the research and the research expenditures by faculty and students was one of the next biggest barriers to Intercollege cooperation. Who is seen to control what is going on was the next but somewhat less important barrier.

Careful and continuous attention is necessary to ensure that students, faculty, departments, colleges and Intercollege units get full credit for their involvement in research, wherever the research is conducted. Credit includes giving appropriate credit for the receipt of grants or contracts, research expenditures, and identification of participants by title and affiliation in both department and intercollege units.

Further, the person responsible for the Intercollege Research Programs should have the opportunity to interact on an equal level with the various other academic administrators with generally similar oversight responsibilities. Specifically for Penn State, the individual with the administrative oversight responsibility for well over a thousand University personnel as well as a third of total University research expenditures should sit on the Council of Academic Deans, better known as the CADs.

Whereas, current membership on CADs includes the Deans or VPs of such academic support organizations as Libraries, International Programs, Enrollment Management and Administration, Outreach and Cooperative Extension, etc. the Director of the IRP is excluded from this opportunity to enhance essential communication and cooperation.

The conclusion can be drawn that intercollege research is viewed as neither academic nor important to the University. I'm sure that some would argue that the membership of the Vice President for Research on CADs adequately represents the interest of the IRP units. However, I doubt that any Dean would claim that the membership of the University Provost on CADs adequately represents the interests of his/her college.

It is also important when recruiting faculty or heads of academic units (department, college or intercollege), that representation on the search committees be constituted with faculty from both departments and related intercollege units, as may be appropriate in the specific case.

During my tenure as IRP Director, and contrary to advice that it would be a mistake to do so, I instituted a quarterly meeting of all IRP Unit Directors, the hosting of which was rotated amongst each of the units. The fear was that such meetings would elevate the expectation of the IRP directors and increase the fear of upsetting the status quo.

At most of those meetings I invited the Associate Deans for Research from each of the related colleges in an effort to enhance communication, cooperation and understanding. Although I think this was generally helpful to communication and cooperation, except for one special case, I'm unaware of reciprocal invitations.

During the last several years of my tenure at the University, I selected a new or hot research topic of interest and set up a seminar wherein any faculty member from any part of the University was given the opportunity in five to ten minutes to inform other faculty members of the University of his/her research and of any specialized facilities or equipment available for use by others. I also took several bus loads of University Park faculty to the Hershey Medical Center where we spent a day hearing of research and facilities available there. A busload of Hershey faculty was then brought to University Park for reciprocal discussions. I recommend that my colleague and new Assistant VP for Research revive this or a similar effort to stimulate increased intercollege research, as I believe it is needed.

Also, I was very pleased to have accomplished after three years of effort to have received approval for a new policy which enabled research faculty who were asked to profess in departments or in the Graduate School (i.e. teach courses, advise students, supervise thesis research, etc.) to be granted a dual professorial title equivalent to the rank of their research faculty title during the year in which they were requested to profess. This policy, known as PS-24, or HR-24 now, enables research faculty to receive credit for professing and adds encouragement for them to want to profess and to be more engaged with students. I hope that this policy is still in use. It was hard fought for.

Research faculty are extremely important assets of the University, but many feel that they are not viewed as first class citizens of the University, which is most unfortunate.

In my mind, tenure does not necessarily constitute first class citizenship or necessarily indicate value to the University or to society.

It took three years of effort to establish this new policy because those opposed to it argued that it would lead to research faculty suing the University for tenure, in spite of the fact, and to the credit of the University lawyers, who indicated that the risk was at most minimal. I'm not aware of any specific suit of this nature.

The knowledge, skills and experience of the research faculty should be mined for their wealth of know how by the other academic programs of the University. They truly are a valuable asset.

In closing, I would repeat that I am obviously of the strong opinion that interdisciplinary or intercollege research can be a valuable complementary asset to a university. It should not be viewed as a threat to the status quo, but as an opportunity.

There is no question that at Penn State the Intercollege Research Programs have been and continue to be extremely valuable assets. Especially in rallying faculty and student interdisciplinary cooperation in undertaking complex research of importance to society. They have been especially successful in attracting industrial involvement through the establishment of industry/government/University research consortia.

The vision of former President Walker and former Vice President Osborn was an excellent one. A vision that has not outlived its value and should not be lost, but which I fear is fading. They correctly saw that the inter-college units were a way to help assist the University to become a major research university, which it has become. A University that is capable of conducting a wide spectrum of research important to society and which prepares students for the world beyond the campus.

The existence of the Intercollege units provides visibility to the University's commitment to research in specific areas that might otherwise be too diffuse or unrecognizable to be effective.

Without question, it was my distinct pleasure to have been directly involved with the intercollege units during their formative years and over a major part of my University career. And it is my pleasure to be invited back to be with you today and to discuss my views on *The Driving Forces for* and *Barriers to Interdisciplinary Research*.

The Local Experience: Interdisciplinarity in a Major Research Universtiy ▶▶

Robert T. McGrath
Associate Vice President for Research,
The Pennsylvania State University

We are fortunate to have present so many distinguished materials research scientists from universities, industries, and governments from around the globe, as well as distinguished scholars representing research policy, research administration, and the interplay between Science, Society, and Technology. My compliments to the conference organizers for bringing together such an impressive collection of experience and expertise.

The case study in interdisciplinary materials research at Penn State dates back into the 1950s. Capitalizing on research collaborations that had evolved, the first Interdisciplinary Graduate Degree Program in Solid State Technology was established at Penn State in 1959. Within a few years, the program had a Ph.D. graduation rate in excess of 20 per year.

In 1962, the underlying interdisciplinary research program in Materials was formally consolidated with the establishment of the Materials Research Laboratory with Professor Rustum Roy as its first director. Since that time, MRL has flourished and expanded its research programs to include ferroelectric materials, dielectric materials, piezoelectric materials, particulate materials, and much more.

Today, under the leadership of Professor Gary Messing, the MRL and its associated graduate program of study in Materials, continue to serve as models for Interdisciplinary Research and Education at Penn State. Their example has led the way for establishment of a multitude of Intercollege Graduate Degree Programs at Penn State, rooted in interdisciplinary research, and administered directly under the Graduate School. At present, Penn State offers Interdisciplinary Graduate degrees in the following fields of study.

Program	Year Established	Current Enrollment
Acoustics	1965	75
Bioengineering	1970	40
Demography	1987	65
Ecology	1972	33
Environmental Pollution Control	1971	70
Genetics	1970	23
Integrative Biosciences	1996	29
Materials	1962	85
Nutrition	1986	61
Operations Research	1991	25
Physiology	1964	35
Plant Physiology	1984	30
Quality and Manufacturing Management	1995	37

These programs of study are closely linked to Penn State's Intercollege Research Programs and emerging Research Consortia.

Intercollege Research Programs
Materials Research Laboratory—Gary Messing, Director
Environmental Resource Research Institute—Archie McDonnell, Director
Institute for Arts & Humanistic Studies—Robert Edwards, Director
Institute for Policy Research & Evaluation—Irwin Feller, Director

Pennsylvania Transportation Institute—Bohdan Kulakowski, Director
Population Research Institute—Mark Hayward, Director

Emerging Research Consortia
Materials Research Institute—Carlo Pantano, Director
Environmental Research Consortium—Archie McDonnell, Director
Life Sciences Consortium—Nina Fedoroff, Director
Children, Youth and Family Consortium—Karen Bierman, Director

The structure of these research and education programs, their interrelationships, and optimization of their contributions to the University's principal missions of Education, Research, and Service, continue to evolve.

I believe that you will hear more about these programs and Penn State's strategic plan for nurturing and expanding these key research thrusts during the presentation by Dr. Rodney Erickson, Penn State's Executive Vice President and Provost.

Throughout the past 40 years, a period witnessing dramatic social changes, technological changes, and changes in sponsorship of basic and applied research, the MRL and the materials research community, all of you, have provided Penn State University and the global materials research community with leadership for navigating the sea of change to the great benefit to the students served and to the continued advancement of materials research.

The conference provides yet another opportunity to share experiences across the spectrum of materials research institutions represented here, and to learn from the successes and from the mistakes that have been made.

At Penn State, we will certainly take advantage of the continue leadership the materials research community, all of you, provides to enhance the productivity and growth of all of the University's Interdisciplinary Research and Education Programs.

The Local Experience: Interdisciplinarity in a Major Research Universtiy ►►

Rodney A. Erickson
Provost, The Pennsylvania State University

I have a strong personal interest in interdisciplinary activity. I was fortunate that my graduate advisers at the University of Minnesota and the University of Washington encouraged all of their students to take courses outside of the discipline. Also, I was privileged for most of my professional career to have a joint appointment in two colleges—business administration and earth and mineral sciences—and to have colleagues who appreciated my interest in interdisciplinary work. Much of my research has been interdisciplinary in nature, relating to urban and regional economic analysis, international trade, and economic development policies. About one-third of my publications were jointly authored with people from other disciplines, and these have been among my publications most often cited by other researchers.

Interdisciplinary approaches are the future direction for research, teaching, and other dimensions of education. As Rustum Roy has written, the real problems of society do not come in discipline-shaped blocks. Disciplines have built silos, and we need to build a lot of halls between these silos.

Major research universities like Penn State interface with society's problems very directly through our research efforts and they do this very differently than they did 10 or 20 years ago. Interdisciplinary units have shown enormous growth in the last 30 years and they are here to stay. At Penn State, interdisciplinary units represent 10 to 15 percent of our research activity, or 30 to 35 percent if we include the Applied Research Lab (ARL).

Strong interdisciplinary programs are based on strong disciplines; we need to have both. At Penn State, we are building on our strong departments and colleges, and we are making investments in five key areas—areas that do not fit in any one discipline. These areas are Life Sciences; Materials Science; Children, Youth, and Families; Environmental Studies; and Information Sciences and Technology.

These initiatives are in addition to Penn State's long history of intercollege research programs (IRP) and intercollege graduate degree programs. We currently have six IRPs, plus the Applied Research Lab. These are major cross-college research efforts supported by the Office of the Vice President for Research. Intercollege graduate degree programs, drawing on the resources of faculty and courses from several colleges, account for nearly 10 percent of the University's graduate enrollment.

To encourage interdisciplinary activity, we need to take a hard look at what the barriers are and then devise ways to minimize the barriers. For example, some departments lay out journal lists—'A' journals, and so forth. I believe in the highest standards of academic excellence, but I find this approach incredibly stifling, and counterproductive in broadening the scope and influence of departments and disciplines. It is more important to publish high quality research that will get wide distribution. It is also important to ask who will be reading the material and what the wider impact will be.

The support and involvement of college deans and department heads is very important. We need leadership in promoting interdisciplinarity from the top levels of the University. We need to have more joint appointments and to use more cross-listing of courses. Faculty can help by encouraging their students to seek out related fields of study.

Universities need to value interdisciplinarity in the resource and reward structure and to examine the effect of promotion and tenure guidelines. Government and foundations, of course, have tremendous leverage over the whole process. Through my experiences with agency funding programs, I have noticed a shift toward more interdisciplinary programming.

We have a lot left to do, as a university, as departments, as foundations, government and industry. Much of the challenge rests with us. It rests with those of us who see the benefits of interdisciplinary activity and who are in a position to support this timely and essential endeavor.

Experience With I³R
in Other Fields

Science-Technology-Society: An Example of Interdisciplinarity in Education ▶▶

Robert E. Yager

Science Education Center, University of Iowa

Science-Technology-Society (STS) is perhaps the most widely institutionalized example of interdisciplinarity in education. It became a major example of the move from a strictly discipline focus for K-12 and collegiate education in the 70's and early 80's. The current movement on STS in K-12 began in Europe with major curriculum efforts in schools underway in the U.K. and the Netherlands.

It was picked up by John Ziman in his book *Teaching and Learning About Science and Society* (1980). Ziman identified several courses and titles and special projects that had many common features. All were concerned with a view of science in a societal context–a kind of curriculum approach designed to make traditional concepts and processes found in typical science and social studies programs more appropriate and relevant to the lives of students.

Currently STS is widely recognized as a major reform effort as correctives are sought around the globe to attain a scientific literacy for all. Typical school science is ineffective in producing students who are knowledgeable of the basic laws and theories known to scientists as accepted views of the workings of nature. Nor are we successful in education in producing students who think that such views of the universe are important and/or relevant to their own lives.

Interestingly, technology (how the human-made world operates) is seen as more important today than science (how the natural world operates). And yet, it is rarely taught to all students across the elementary and middle school years and often only to the non-college students in high school. Each attempt at reform of science education in the United States over the past 200 years has moved science instruction to the practical, that is, science that could affect the reform trend was the one that characterized the national efforts during the 1960's–following the soviet launch of Sputnik in 1957. Although the artificial satellite was more of a technological achievement than a science one, attention and funding were directed toward reform that illustrated and emphasized basic science–that is, the science known and practiced by scientists. Scientists in their disciplines were instrumental in determining the central themes for physics, chemistry, biology, and earth science. The science disciplines were moved to the junior high school curriculum–a replication of the discipline bound programs of the high school. Process skills used by scientists became a central focus for science programs for the elementary school.

Reform was seen as a return to basic science, especially concepts and themes currently accepted by scientists. Reform also ushered in a renewed look at the process skills employed by scientists–an effort first popularized as a second dimension for the focus for school programs in the 1930's. Jerrold Zacharias, the architect of the Physical Science Curriculum Study (the one reform underway before 1957), identified the axiom that defined the science reforms of the 1960's and early 1970's. He said, "Science, when presented in a way known to scientists, will be inherently interesting and appropriate for all learners."

Disillusionment with these reforms following Sputnik was widespread by the mid-1970's. Many blamed the social problems, including the Vietnam conflict, on science and technology. Nearly all social institutions–including schools–were under attack.

Many efforts of the late 1970's and early 1980's resulted in a database that was useful as new reforms were conceived. th3e economic woes in the U.S. hastened a fundamental change in United States policy that called for renewed efforts and funding for improving school science, technology, and mathematics. The appearance of technology was a major shift–something that is the connector for STS.

Many were still convinced that the efforts of the 1960's were correct and that we had merely abandoned them too soon. The controversy led to a focus on new research to find out more about the learning process. In 1983 funds were awarded to study physics and engineering majors at colleges and universities–with the simplistic notion that this information would help educators know how to deal with students in schools with less interest and expertise in science.

Surprisingly, however, the research found that these most interested and successful students had not learned. They could not use the information (concepts) they seemed to know in classes and on examinations. Nor could they use the process skills they "learned" and practiced in laboratories. It was soon apparent that the reform efforts for the 1980's would have to start in ways neither generally conceived nor tried in the past.

This was the national situation as interest and trial with STS in science education was borne. The promise of STS was first seen in trials in other countries and circumstances. Such new experiments with reform internationally provided the foundation for STS experiments across the United States. Recently efforts have been undertaken to assemble research results concerning the STS movement in the United States with some consideration of continuing initiatives the world over. (Yager, R.E., 1996.) This collection of results with STS, 1985-1997, remains the most complete review of results with interdisciplinarity in K-12 science.

One of the major problems with any reform effort is visualizing all its features. STS efforts in the United. States were so extensive by 1988 that LaMoine Motz, then president of the National Science Teachers Association (NSTA) appointed a Task Force to offer a definition of STS for NSTA–the largest professional society for science education. In 1991 the recommendations of this Task Force were unanimously adopted by NSTA. The essence of the definition is captured by the statement: STS is the teaching and *learning* of science-technology in the context of human experience. The report ended with the following statement: "STS requires that we rethink, restructure, reorganize, rewrite, and revise current materials (i.e., curriculum, texts, audiovisuals) used to teach science. STS will require a realignment of goals and objectives and a reallocation of resources. STS will require re-education on all levels from policy makers to teachers to parents. Such reform of science education is essential." (NSTA Handbook 1998-1999, p. 229)

STS in education continued to be attacked by discipline-bound educators in high schools and undergraduate science departments. However, others have charged that the only places that the science disciplines exist are in high schools and colleges in terms of curricula and courses. Most research in science is now inter-disciplinary requiring information, skills, and new technology in order to deal with the real questions of our time. The debate is whether learning can occur without the language and the skills as traditionally taught in secondary schools and colleges as separate disciplines. The research efforts certainly favor the inter-disciplinary approach. This means dealing with real questions and problems as seen by students and others opposed to simple questions that arise from textbooks and instructors. Many argue that dealing with current problems and issues are the best preparation for the future.

STS forces all to consider the context or the situation that creates a desire for students to learn. STS provides an example of Perrone's (1994) elaboration of eight ways in which the minds of students become engaged. These eight facets include:

1) Students help define the content.

2) Students have time to wonder and to find a particular direction that interests them.

3) Topics have a "strange" quality–something common seen in a new way, evoking a "lingering question".

4) Teachers permit–even encourage–¬ different forms of expression and respect students' views.

5) Teachers are passionate about their work. The richest activities are those "invented" by teachers and their students.

6) Students create original and public products: they gain some form of "expertness".

7) Students *do* something–e.g., participate in a political action, write a letter to the editor, work with the homeless.

8) Students sense that the results of their work are not predetermined of fully predictable.

Once intellectual engagement is realized, learning is likely to occur. Since real learning rarely occurs as a result of typical instruction characterizing most K-12 and college science classrooms, new approaches are needed (Lochhead & Yager, 1996; Mestre & Lochhead, 1990; Brooks & Brooks, 1993). Typical teaching places premiums on students who can repeat, recall, and regurgitate science concepts and process skills that provide the curriculum frameworks in most settings. As late as 1990, Mestre and Lochhead proclaimed that the goals for science education can be defined in terms of two dimensions, namely concepts and processes (Mestre & Lochhead, 1990). But this two-dimensional view of science is not adequate for the current reform initiatives; these two represent only part of what the new *National Science Education Standards* (NRC, 1996) outline as essential science content. For example, technology, science for meeting personal and societal challenges, and the use of science concepts and process skills are identified as valid forms of content for science. But more important than science content is the issue of how teachers teach and what students are expected to learn.

Reinsmith (1993) has delved more deeply into real learning–that which goes beyond repeating, recalling, and regurgitating (perhaps the 3Rs so many still equate with evidence that learning has occurred). He has described learning in the following ways:

1) Learning first takes place by a process much like *osmosis*.
2) Authentic learning comes through the trial and error.
3) Students will learn only what they have some proclivity for or interest in.
4) No one will formally learn something unless he/she *believes* he/she can learn it.
5) Learning cannot take place outside an appropriate context.
6) Real learning connotes use.
7) No one knows how a learning moves from imitation to intrinsic ownership, from external modeling to internalization and competence.
8) The more learning is like play, the more absorbing it will be.
9) For authentic learning to happen, time should occasionally be wasted, tangents pursued, side-shoots followed up.
10) Traditional tests are very poor indicators of whether an individual has really learned something.

STS teachers must be reflective and must think of their teaching as s science itself. They must question their behaviors and actions and hypothesize about how they might impact learning both negatively and positively. They must raise questions for which possible answers can provide the basis for observations, data collection, and data analysis to determine the validity of the idea (original explanation) that was proposed. This collection of evidence concerning the validity of a personally constructed explanation exemplifies science, constructivist thinking, and the teaching actions required in STS classrooms. The specific procedures have been defined and examples provided in two publications that include research data about success with STS in education, namely, *Science/Technology/Society as Reform in Science Education* (Yager, 1996) and *NSTA What Research Says About STS*, vol. 7 (Yager, 1993).

STS teaching strategies illustrate how the approach is an example of interdisciplinarity. Nineteen features characterize STS and illustrate how and why discipline formats do not work. STS results in teachers who:

1) Encourage students to identify and initiate their ideas/problems/issues
2) Accept a variety of student responses
3) Allow adequate time for student expression and analysis
4) use student ideas to drive lessons
5) Seek elaboration of initial Reponses offered by students
6) use various local resources (people, places, and situations)
7) Focus on current societal issues related to science and technology (i.e., issue-oriented science)
8) Encourage students to explore and use technology in the learning process
9) Ask higher-order thinking questions
10) Give students opportunities to experience applying their knowledge in meeting everyday challenges
11) Create situations leading to increase career awareness related to science and technology
12) Motivate students to take actions that illustrate exemplary citizenship roles
13) Work with students in the process of their individual learning
14) Encourage student-student verbal interactions
15) Encourage students to use higher-order thinking skills
16) Use a wide variety of assessment tools as well as those arising from student self-evaluation

17) Design themes that promote questions that illustrate science as fields of inquiry (without discipline boundaries)
18) Change the curriculum to utilize student present understanding
19) Use varied methods of teaching

Opponents of STS fear that students will not learn the basic constructs and the skills, which have characterized the traditional disciplines of science. And yet traditional teaching does not result in real learning of such concepts either (at least for 85% of the students). This was illustrated in the 80's when major funding was awarded to cognitive scientists to unravel the workings of the brain in an effort to help educators do a better job in producing students who had learned. The studies first undertaken were to look at the best college students—those majoring in physics and engineering. The researchers gave undergraduate students tough problems and sought to study the ways they solved them. Unfortunately, however, the research team established instead that most did not understand and could not solve the problems. The engineers were worse where 90% of the undergraduates studied could not solve the real-world problems given. They could only repeat, recite, recall what they had studies. They could not use the information and/or the skills that they could demonstrate in the laboratories and on course examinations in new situations.

The cognitive science research has provided rich evidence for the necessity of STS in education and more universal moves to interdisciplinary approaches. The National Science Education Standards focus on the importance of changing to a consideration of how we teach as opposed to what we teach. Context for learning is more important than the concepts and skills we want to teach. These needed changes in teaching approach are outlined in the Standards as follows:

LESS EMPHASIS	MORE EMPHASIS
• Treating all students alike and responding to the group as a whole	• Understanding and responding to individual students' interest, strengths, experiences, and needs
• Rigidly following curriculum	• Selecting and adapting curriculum
• Focusing on student acquisition of information	• Focusing on student understanding and use of scientific knowledge, ideas, and inquiry processes
• Presenting scientific knowledge through lecture, text, and demonstration	• Guiding students in active and extended scientific inquiry
• Asking for recitation of acquired knowledge	• Providing opportunities for scientific discussion and debate among students
• Testing students of factual information at the end of the unit or chapter	• Continuously assessing student understanding
• Maintaining responsibility and authority	• Sharing responsibility for learning with students
• Supporting competition	• Supporting a classroom community with cooperation, shared responsibility, and respect
• Working alone	• Working with other teachers to enhance the science program

STS forces teachers and students to look for evidence of understanding and for models for developing real meaning of science and parallels from the history of science Evidence is sought for establishing that both science and technology are th4e subjects of inquiry and study. The focus must be upon question generation and use. Just as the use of science skills and concepts are important, evidence needs to be sought to indicate student mind engagement has been achieved and maintained (Perrone, 1994) and that real learning is in evidence (Reinsmith, 1993). This should be of concern in all facets of instruction, both in education and science. The views of constructivist teaching practices arise from the interpretations and ideas from Brooks & Brooks (1993) and Yager (1991). A constructivist teacher is one who:

1) Encourages and accepts student autonomy, initiation, and leadership
2) Allows student thinking to drive lessons: shifts content and instructional strategy based on student responses
3) Asks students to elaborate on their responses
4) Allows wait time after asking questions
5) Encourages students to interact with each other and with the teacher
6) Asks thoughtful, open-ended questions
7) Encourages students to reflect on experiences and predict future outcomes
8) Asks students to articulate their theories about concepts before presenting your understanding of concepts
9) Looks for students' alternative concepts and designs lessons to address any misconceptions

Richard Feynman, Nobel laureate, once described three sources for the real content of science. The first of these was a focus on the things we know we don't know, i.e., the questions about the natural world for which we do not have answers–but for which we strive to get answers. Feynman indicated, however, that this was the smallest and easiest part of science. A second source of science content was a focus on the constructs of nature, which are generally, accepted by the academy that are wrong (i.e., the things that "we know that aren't so"). And, the third area for science content according the Feynman was described as "the vast infinity of ignorance about things that we don't even know that we don't know." Such views of science "content" illustrate the power of necessity of inter-disciplinary approaches to education.

This reminds many of us in education of the ancient Tao of 3000 years ago when the Master saw his job as not teaching people to know–but instead to teach them that they did not know. "For if the people know they do not know, they will find their way. People who know are difficult to guide." This may be our problem today. We have too many people who know too much about too little and see their primary job as sharing their knowing with students. And yet we no know that meaning of one human being can not really be shared. Meaning occurs only when each individual constructs in his/her own brain for him/herself. Interdisciplinarity allows us to focus on real questions/problems–the starting points for science itself. Disciplinarity forces us into boxes which do not encourage learning–only the "pretending" to know.

Interdisciplinarity in the Social Sciences ▶▶

Irwin Feller

Director, Institute for Policy Research and Evaluation,
and Professor of Economics
The Pennsylvania State University

Why is there a need, as at this workshop, to once again promote the case for interdisciplinarity? Is it not true that interdisciplinary research (and education) is vocally championed by university presidents, industrial representatives, and government agencies? Doesn't it correspond to the internal logic of the contours of scientific problems, advances in scientific equipment and instrumentation, and the applicability of academic research to broad swathes of a nation's intellectual, economic, and societal needs?

The answer in part rests in understanding that interdisciplinarity has two overlapping but distinct implications. On the one hand, it entails bridge-building across traditional academic structures. On the other hand, it can entail the restructuring of a university's organization. Numerous, almost stock, impediments exist to the former (e.g., distribution of indirect cost recovery funds); these typically can be resolved with reasonable institutional policies and an overarching commitment to the benefits generated by interdisciplinary research and activities. Restructuring, which may mean creating new colleges, divisions, or intercollege research units, can be more threatening to budgets and the control of resources, including the commonplace (if condescending) phrase, "their faculty," by college-level academic administrators.

Related to the bridge-building and restructuring concepts, interdisciplinarity's place on a university campus is shaped by two distinct influences.

One is grounded in disciplinary breadth and thus is a function of the intellectual openness and life cycle productivity of a field's faculty. The other is based in university policies and organization, and the priorities of academic administrators. Thus, while it is not without formidable challenge, it is (relatively) more amenable to change in the short run.

The first, a traditional but nevertheless valid response, is rooted in the sociology and economics of science. "Good scientists," observes Medawar, "study the most important problems they think they can solve." Disciplines differ in the extent to which their most important problems are defined within or at the boundaries of existing theoretical frameworks, methodologies, or equipment. Much of the intellectual ferment in the life sciences and engineering exists at the boundaries of conventional academic departments. In effect, researchers are saying that the frontiers of science are located somewhere within the regions created by the intersection of disciplines. Therefore, performing cutting-edge research requires collaboration with faculties in other disciplines. Supported by federal agency programs such as Engineering Research Centers and Science and Technology Centers, which embody the same scientific worldview, these fields are alive with interdisciplinary research projects. Other disciplines, such as chemistry and economics, are described by university administrators as comfortable within their borders and less likely to be involved in interdisciplinary collaborations.

An economic perspective on the structure of science also supports a disciplinary orientation. Discipline-based research, according to Stigler, is the efficient means of generating new knowledge. He writes that "…specialism is the royal road to efficiency in intellectual as in economic life." A corollary to this statement is that it "is impertinent and unhelpful for outsiders" to "assert that the problems which are dealt with by specialists are less helpful than other problems."

Institutional policies also manifestly affect the viability of interdisciplinary activities on a campus. Here the setting is more complex, involving increasing tensions between faculty (and sponsor) interests in interdisciplinary undertakings and administrative ukases to constrain these interests within established jurisdictional bounds. The university setting also is schizophrenic: university leaders simultaneously note scientific trends and national interests that require interdisciplinarity, while at the same time adopting strategic planning, benchmarking, and performance measurement practices that allow little breathing room for interdisciplinary work. As noted by the Government-University-Industry Research Roundtable, "Interdisciplinary programs are orphans within the fiscal bureaucracy of the university." To these enervating practices can be added a myriad of policies and practices, such as disputes about the apportionment of indirect cost recovery funds, which impede interdisciplinarity.

Most importantly, universities have failed to reconcile the rhetoric emanating from the president's office with the actions of deans and department heads. Interdisciplinary themes may be sounded from academic heights, but faculty most clearly hear and attend to messages about decision making relating to promotion, tenure, salary, and departmental and faculty resources in their villages. A recurrent theme voiced on several research university campuses is the power of deans to throttle interdisciplinary initiatives.

Interdisciplinarity will reach its natural limits—the limits shaped by the intellectual and scholarly interests of faculty, the interest of sponsors, and society at large—only when university rhetorical commitments to interdisciplinarity permeate administrative levels. Achieving this requires, at a minimum, appointment and retention of academic administrators who, if not committed to interdisciplinarity, at least do not stand in its way; development of performance metrics that reward collaboration across academic units; recognition of the scholarly merit of collaborative work and the journals in which interdisciplinary research is published; and institutional resources to support interdisciplinary initiatives. Faculty must be allowed to pursue their research interests, to collaborate with faculty in other units, and to conduct their research in the organizational setting y find most productive.

The Integrative Imperative:
Interdisciplinarity in Medicine ▶▶

Andrew T. Weil
Tracy W. Gaudet
Program in Integrative Medicine
University of Arizona

Our society is in the midst of a heated debate about national health care policy, primarily focused on who is going to pay the bills. These arguments obscure the more critical issues: conventional medicine has become so enamored with technology that it has reached an impasse where no one can keep up with the high costs, and it is unable to deliver the kinds of healthcare that more and more people demand. A survey by Eisenberg et al in 1990 revealed that 34% of people surveyed had used at least one unconventional therapy in the previous year. When extrapolated to the U.S. population, this data suggests that in 1990 there were 425 million visits to alternative medicine practitioners compared to 388 million primary care physician visits. Most importantly, this survey revealed that of those patients using alternative medicine, 72% did not inform their physicians. NEJM 1993;328:248-252 A 1997 survey (n=1500) revealed that 42% of individuals surveyed used alternative medicine in the past year, and 74% of those used alternative approaches along with conventional medicine. *The Landmark Report-independently commissioned report*

Government, insurance companies, and managed care organizations are responding to consumer demand as well. The NIH Office of Alternative Medicine was mandated by Congress in 1991 and launched in 1992 with an annual budget of $2 million. The budget this year has been increased to $50 million, and the status upgraded to a Center. In Washington state, the Seattle King County Clinic is the country's first publicly funded "holistic clinic." Increasing numbers of states are mandating reimbursement for alternative providers and more and more numbers of insurance carriers are offering coverage. *Am J Health Promot 1997; 12 (2) :212-222.*

The reasons for this change are many, but can largely be attributed to consumer demand. Consumers of health care increasingly want doctors who have the time and interest to ask and answer questions, who can offer treatments other than drugs and surgery, who value nutritional and lifestyle influences on health and illness, and who understand the natural potential of the human organism for self-repair and healing. Physicians should be trained to understand the subtle and complex interactions of mind, body, spirit, and community and know how to interpret them in health and disease. Physicians need to recognize the primacy of instructing people about how not to get sick, with the treatment of disease as secondary. Physicians should have a critical understanding of a whole range of alternative and natural therapies not taught in medical schools, so that patients are not left in the precarious position of deciding between alternative and conventional treatments with no clinician to guide their decisions. Currently there is no standardized way for physicians to learn this type of medicine.

The fact that physicians are not currently educated about alternative systems of medicine or how to critically counsel patients about their use creates dangerous circumstances. Such a situation is ripe for fraudulent practices and products and bad medical outcomes, and frequently leaves the patient not consulting the physician about choices and instead turning to the health food clerk for medical advice. This is not in the best interest of good healthcare in this country. The medical profession has a responsibility to address this situation by committing to research in these areas and setting appropriate standards for credentialing educational programs as well as practitioners who claim to have expertise in these areas. Most importantly, academic medicine has a responsibility to rigorously educate medical students and physicians about what is known about these systems and what can be offered safely to patients and what practices are not effective or harmful. Physicians should be taught how to intelligently and critically synthesize the best ideas and practices of all systems of treatment.

As a first step toward meeting this need, The University of Arizona College of Medicine has established a **Program in Integrative Medicine**. *Integrative Medicine* shifts the orientation of medicine to one of healing rather than disease, engaging the mind, spirit, and community as well as the body. The integrative approach is based on a partnership of patient and practitioner within which conventional and alternative modalities are used to stimulate the body's innate healing potential. It neither rejects conventional medicine nor uncritically accepts alternative practices. The Program was initiated as a section within the Department of Medicine under the direction of Joseph S. Alpert, M.D., Head, Department of Medicine. The Program is subject to the same guidelines, rules, responsibilities, and standards as other training programs in the College of Medicine.

The Program in Integrative Medicine
The Program in Integrative Medicine is responding to a growing demand for a standardized curriculum of training in healing-oriented medicine from an accredited academic institution. As patients and physicians become increasingly frustrated by the limits of conventional medicine and more interested in alternative approaches, it has become clear that there is no comprehensive way for physicians to be trained in these areas.

The Program in Integrative Medicine offers two main courses of study. The two-year fellowship is designed to train leaders in this new discipline of medicine. The continuing professional education track offers a variety of educational offerings designed for physicians and other health care providers who want to further their knowledge and offer a broader range of services to their patients. Included in these offerings is a two-year comprehensive nonresidential program. Both curricula are designed to be models for similar programs at other institutions. In addition, there is an Integrative Medicine Clinic which creates a model of clinical care within a university setting where patients can be seen for a variety of needs ranging from optimization of health to terminal cancer care. Research is also a high priority for the Program as it seeks to establish the benefits of an integrative approach to medicine and investigates which individual approaches are effective and which are not.

Fellowship Program

The centerpiece of the Program is a two-year postgraduate fellowship. The intent of the Fellowship is to produce leaders in this new field. It is anticipated that after completing the program, Fellows will direct similar programs at other institutions, and consult with local and national organizations to set policy and direction for health care in the 21st century. The educational experience includes a core curriculum of thirteen subject areas, research, and on-going training in the clinic.

One premise of Integrative Medicine is that physicians should participate in their own health and healing. Consequently, personal health as well as intellectual and practical skills are emphasized. To aid in personal progress, Fellows are provided with instruction in nutrition, meditation, exercise, relaxation and other aspects of healthy lifestyle. They are encouraged to use their own bodies as laboratories in this process.

The goal of this curriculum is not simply to add content to existing medical education, but to change the process by which learning takes place. The fellowship experience begins with time for reflection and developing an appreciation for the fact that ours is only one way of looking at health and illness. Emphasis is placed on this process for both the faculty and the fellows throughout the two-year curriculum.

Core Curriculum

Fellows study a core curriculum of thirteen subject areas: Healing-Oriented Medicine, Philosophy of Science, Art of Medicine, Medicine and Culture, Research Education, Mind-Body Medicine, Spirituality and Medicine, Nutritional Medicine, Botanical Medicine, Energy Medicine, Lifestyle Medicine, Complementary and Alternative Medicine, and Medicine, Leadership and Society. Faculty from The University of Arizona colleges of Medicine and Pharmacy, and various undergraduate departments, as well as visiting faculty who are leaders in their respective fields provide the didactic instruction.

Clinical Education

In addition to the core curriculum, much of the clinical teaching takes place in the Integrative Medicine Clinic. Each patient is discussed in a patient conference which includes the Fellows, the medical director, the director, a pharmacist, a nutritionist, and clinicians representing various alternative modalities including manual medicine, oriental medicine, homeopathy, and mind-body medicine. In this forum, the patient's individualized treatment plan is designed. The Fellows screen for problems that require immediate allopathic intervention before any alternative approach is considered. They then learn the art of combining alternative methods with conventional treatment in ways appropriate to individual diseases and individual patients.

The Fellows learn the strengths and weaknesses of various systems of alternative medicine, as well as a working knowledge of select alternative modalities. All Fellows are required to master the theory and practice of guided imagery, and develop a working knowledge in complementary and alternative modalities such as oriental medicine, homeopathy, and osteopathic manipulative therapy. The teaching of these systems is provided both as a part of the core curriculum and on an ongoing basis in the clinic. Fellows may choose to explore any of these modalities in greater depth in the second year or work with other systems of healing contingent on the approval of Program directors and availability of preceptors.

Research Experience

Fellows are also required to be a part of designing and conducting a research project during the course of the fellowship. As a first step, they master critical thinking about research including; how to assess existing research and evaluate its validity and significance, how to formulate valuable questions, and how to design experiments and methodologies that effectively address these questions. Either basic science or clinical research satisfies this requirement of the Fellowship. It is expected that this work will result in papers to be published in peer-reviewed journals.

Fellowship Program Eligibility

The Fellowship program offers four positions per year. Fellowships will be offered only to M.D.s and D.O.s who have completed residencies in primary care specialties.

Integrative Medicine Clinic

Fellows staff an outpatient Integrative Medicine Clinic within The University of Arizona Health Sciences Center (AHSC), under the supervision of the director, the medical director, and other AHSC faculty members. Fellows observe and work with practitioners of a variety of modalities not typically available at allopathic institutions, which also participate in the previously mentioned multidisciplinary patient conference, Fellows present their evaluation of patients, design treatment plans, and track outcomes. Many patients are self-referred to this Clinic; others are referred by physicians in the community and by AHSC physicians.

Emphasis is placed on establishment of rapport with patients; the efficient taking of a history that includes the emotional, psychological, and spiritual aspects of patients' lives; careful listening; assessment of patients' belief systems; and presentation of treatments in ways that increase the likelihood of successful outcomes. The clinic also provides the opportunity for electives for fourth year medical students.

The Research Program

In addition to individual fellow research projects, the Program is developing a long-term research agenda. The goals are three fold. First, the Program seeks to evaluate the efficacy and cost-effectiveness of an integrative approach to medicine when compared to typical conventional approaches. This agenda clearly recognizes that Integrative Medicine is more than a compilation of alternative systems of healing and that the impact of this different approach to the patient and to medicine needs to be investigated. Second, it is essential to evaluate the effectiveness of specific alternative modalities to determine which alternatives should be incorporated into medicine and which should not. Many of these studies will be done collaboratively with other departments and institutions. Third, the Program promotes the implementation of research methodologies that can most effectively address the questions being posed in this emerging field. Just as new science (such as quantum physics) requires new frameworks in which to be understood (such as advanced math), Integrative Medicine may require new methodologies to be quantified. Conventional research methodologies may not always be the most appropriate in addressing certain aspects of Integrative Medicine. The design of this agenda draws on national and local expertise.

Professional Development and Continuing Medical Education

Professional development and continuing medical education are another aspect of the Program in Integrative Medicine. Increasing numbers of physicians trained in allopathic medicine are dissatisfied with the limitations of their current practices and are eager to broaden their base of practice to include sensible and cost-effective alternative therapies.

The Program offers a variety of professional development and continuing education activities designed to meet the needs of physicians and professional health care providers in clinical practice including nurses, pharmacists and other allied health professionals. The content parallels the offerings of the Fellowship program with activities tailored to meet the needs of individual practitioners. To date the Program has offered: a quarterly mini-conference series introducing topics in Integrative Medicine. These three hour sessions have been held at the College of Medicine and satellite broadcast to remote sites across the country; one-day intensive courses which provide hands-on exposure to select modalities; week-long, comprehensive conferences designed for physicians in practice. In collaboration with other academic institutions, the program will be offering week-long subspecialty courses in such areas as cardiology, oncology, and psychiatry. Recognizing the time constraints of physicians with active practices, the Program is committed to developing long distance learning opportunities through innovative uses of technology such as online courses and teleconferencing. The Program intends to offer a comprehensive certification course for practicing physicians paralleling the fellowship. This course of study will use a combination of on-site and distance-learning models.

Program Support
The Program is within the Department of Medicine and enjoys the full support of James Dalen, M.D., M.P.H., Dean, College of Medicine and Joseph S. Alpert, M.D., Head, Department of Medicine. This degree of support within a major medical institution is unique and allows for an unparalleled opportunity at The University of Arizona.

Looking Toward the Future
The intent of the Program is to influence the future of medicine by redesigning medical education and establishing Integrative Medicine in academic medicine. One of the key objectives is to implement the

philosophies and practices of Integrative Medicine across all of healthcare, rather than establishing a new medical subspecialty. As the Program at The University of Arizona evolves, many educational opportunities will emerge such as the development of:

- Integrative Medicine education as a part of the standard medical school curriculum and within all residency and fellowship programs. Further development of the national collaboration of academic institutions committed to furthering Integrative Medicine
- Standardized certification course for practicing physicians and other health care professionals
- A national network of Integrative Medicine practitioners
- These efforts begin to address the needs that the emergence of this new field of medicine create. There is a need to demonstrate to managed care organizations, insurance companies, and government agencies, as well as to health care providers and patients, the efficacy and cost-effectiveness of integrative approaches. In so doing, integrative philosophies and therapies previously thought "unconventional" will be incorporated into mainstream medical practices and pave the road to new paradigms of healing for the 21st century.

Curriculum Topics:

Healing-Oriented Medicine
- The nature of the body's healing system
- Case studies of spontaneous healing
- The placebo response as a therapeutic ally
- Strategies for protecting, enhancing, and activating the healing system

Philosophy of Science
- What science is and is not
- How quantum theory and chaos theory supersede old Newtonian/Cartesian models

- The relationship between medicine and science
- The scientific basis of treatment modalities

Art of Medicine
- Effective communication and the art of suggestion
- Relationship centered care
- The role of intuition
- Matching therapeutic approaches with individual patients
- Engaging one's own healing process

Medicine and Culture
- History of medicine: the origins and development of such major systems as ayurveda, traditional Chinese medicine, homeopathy, osteopathy, and naturopathy in addition to allopathic medicine
- Cultural influences on medical thinking
- Cultural definitions of health and illness
- The role of ceremony and ritual
- Anthropological perspectives

Research Education
- How to design and ask questions
- How to design quality studies
- How to interpret research findings
- How to identify research methodologies appropriate for Integrative Medicine

Mind/Body Medicine
- The scientific basis of mind/body interactions
- Identification of illnesses with prominent mind/body components
- Critical review of mind/body therapies
- How to assess the moods and belief systems of patients

Spirituality and Medicine
- The spiritual dimension of human life and its relevance to health, illness, and healing
- Distinction between spirituality and religion
- Spirituality and medical outcomes
- Spiritual healing techniques, including the power of prayer

Nutritional Medicine
- The contributions of diet to health and illness
- Specific information on benefits and risks of common foods and dietary patterns
- The therapeutic role of vitamins, minerals, and other nutritional supplements
- Nutritional therapies for specific diseases

Botanical Medicine
- Rationale for using plants as therapies
- Medical plants: their preparations, uses, benefits and risks, and possible interactions with pharmaceutical drugs
- Identification of a basic repertory of botanical remedies with known safety and efficacy

Energy Medicine
- History and critical review of energy-based modalities such as qi gong, jin shin jizutsu, therapeutic touch
- Health risks and benefits of various forms of energy
- Scientific basis for subtle energies

Lifestyle Medicine
- Effect of lifestyle choices including smoking, vitamins, exercise, alcohol, food
- Techniques for motivating behavior change

- Physician as role model in addressing health and healing
- Health and the environment

Complementary and Alternative Modalities
- Distinction between alternative and integrative medicine
- History of Complementary and Alternative Medicine (CAM)
- Scientific basis and critical review of CAM
- Practical knowledge of CAM including:
  - Guided Imagery and Hypnotherapy
  - Acupuncture and Oriental Medicine
  - Homeopathy
  - Manual Medicine

Leadership, Medicine, and Society
- Leadership and the facilitation of social change
- Business planning and management skills
- Political, legal, and ethical aspects of Integrative Medicine
- Physician as teacher
- Collaborative relationships among health professions

Comments
and
Recommendations

Comments and Recommendations
by Participants ▶▶

The following recommendations (and comments) were submitted by individuals. Some of the main ideas are included in the Executive Summary. Their statements contain unique nuggets of ideas for supporters and performers of such interactive research. They are presented without editing, except highlighting of key suggestions, in alphabetical order of the author's name.

****

Arden L. Bement, Jr.
Department Chair, Nuclear Engineering, Purdue University
Former Chair NRC Committee on Benchmarking Material Science

- There is often confusion between the terms, *"multi-investigator"* and *"interdisciplinary"* research among the various federal agencies. While it is quite common for investigators from different disciplines to be funded under a single grant, this may not involve interdisciplinary research. True interdisciplinary research requires close collaboration and synergy among the investigators in addressing a common problem...*and this is rarely achieved.* Funding agencies should be sensitive about the difficulty and special *institutional incentives* that may be required to achieve the full intellectual leverage possible from interdisciplinary research.

- Among the success factors involved in successful interdisciplinary research, the role of the *charismatic leader*, who can identify the interesting challenging problems to be solved and then organize meaningful collaborations to address these problems is paramount. However, providing incentives to allow potential leaders to surface and thrive is often a daunting challenge for universities with limited discretionary resources. Innovative approaches to *cultivate younger faculty members who show high potential for leadership* is an important consideration for both funding agencies and universities.

- Outstanding interdisciplinary research programs exist among collaborators throughout the U.S. and around the world. *True collaborators usually find ways to circumvent geographical, funding and bureaucratic constraints.* Research into the ways in which the more effective of these efforts are organized and managed, especially with the use of modern communications technologies, would be a valuable investment to assist R&D planners and decision makers.

****

Richard Brook

Chief Executive, Engineering and Physical Sciences Research Council
London, England

A paradox in research support is that all scientists claim to be interdisciplinary and that all funding agencies strive to be interdisciplinary and yet the ancient barriers persist. One major contributor to the paradox is the conflict between evaluation—research quality assessment-on the one side and the relaxing of disciplinary loyalties on the other.

Peer review is almost universally used to select future research projects. It is increasingly used to evaluate past performance as a basis for continued research support. In both instances a CONSENSUS process is asked to determine ORIGINALITY, i.e., departure from consensus. The consequence is not surprising: the peer selection tends to reproduce the characteristics (age, gender balance, institutional affiliation, disciplinary allegiance) of the peer community. This is less damaging for that variety of interdisciplinarity (materials science?) which is simply a redefining of disciplinary boundaries between one generation of research and the next; it is life-threatening for that variety which reflects an original and untried mix of skills brought together to explore an uncharted research domain. When the domain is obvious (bioinformatics, climate change response) then specific programs can be set in place; when the domain remains ill-defined, peer review can be trusted to be unimpressed.

Is this a problem? There is one view that proposals and reviewing are conducted in a code language between applicant and reviewer and that the work as eventually performed can take risks which remain unmentioned in the proposal. There is another view that private agencies can relax the peer review constraints of public sources. These are, however, sidelining the problem more than solving it.

Therefore it becomes attractive to work on the peer review itself. Possible avenues include: -peer reviewer engineering. Select for the reviewers those who have the characteristics of the research community which it is wished to create; -use US-style program managers. Allow a particular viewpoint to determine the discipline set (dangerous for single agency systems!); -introduce CONTENTION. Genuine originality promotes debate; but how is this to be creatively staged in a committee culture?

****

George Bugliarello
Former President, Polytechnic University of New York
Founding Editor of Technology and Society, Editor of NAE Bridge

- *Fostering inter- and multi-disciplinarity is a very powerful mechanism for the development of knowledge.* It needs to receive much more sustained attention than is the case today.

- The fostering *needs to be conceived broadly*, as spanning not only disciplines, but also institutions and institutional components, across universities, colleges, departments, as well as reaching from universities to industry, from universities to schools, and *vice versa*.
- *Interdisciplinarity*—the blending of disciplines *is intrinsically more difficult than multidisciplinarity*. But both, if they are to blossom, require support budgetarily, organizationally, sociologically, psychologically, from administrations, from government, from foundations.
- *With support comes responsibility*—identification of realistic objectives for the inter- or multi-disciplinary efforts, and assessment of achievements and of *expected impacts on knowledge and society*.

****

Robert Cahn

Department of Materials Science & Metallurgy
University of Cambridge England
Founding Editor of J. Materials Science

1) *In education:*

- To teach properly, the teachers must spend much time on *research in a variety of topics*. A single large, highly specialized research unit is apt to be fatal.
- Some way needs to be found of requiring even undergraduates *to read some original papers (preferably mutually contradictory ones), and to comment critically on them.* My experiment at Sussex University in introducing 'Critical Literature Surveys', for credit, was one approach. There is no other way to convince students that scientific knowledge is never cut and dried.
- *Linking themes need to be searched out.* (MIT has recently been experimenting with this). One example: derivation of mathematical random-walk theory, with modifications, and application to rubber

elasticity in polymers, diffusion in crystalline inorganic materials. Another: point defects in crystals, related to diffusion (again), age-hardening kinetics (non-equilibrium vacancies are involved), equilibrium and non-equilibrium segregation (vacancy wind), etc.

- To enhance the sense of a discipline constructed out of an array of contributing disciplines, *central themes need to be emphasized*: in MSE, this would be a study of the evolution and nature of microstructure, stereology, a broad survey of characterization (including the analysis and role of trace elements over a very wide range).

2) In free-standing research units:

- If this is in a university, then those working in the participating disciplines need to be able to look to relevant academic departments for stimulus and support. The MRLs did this superbly. The lack of a metallurgy department (or even an engineering school) killed the original concept of the Institute for the Study of Metals at Chicago University and turned it into something quite different.

- If not in a university, *then organized contact between participating kinds of specialist* is important. In science, we don't have an inverse square law of interaction at a distance…we have an inverse cube law at least. The old-fashioned British tearoom is valuable, so long as people are discouraged from segregating to separate tables …but that isn't easy. Colloquia/*seminars which people are firmly expected to attend*, and in which speakers temper the wind to the shorn lamb, are crucial. In my small department of materials science at Sussex University, I insisted on *running weekly seminars* from the very beginning. *A significant number of these must be by internal speakers*. In such a lab, unlike in an academic undergraduate school, a specialized objective can be an advantage, so long as many skills contribute to its realization.

3.) Some Suggestions for Agencies:

- Other things being more or less equal, *favor proposals from two or more proposers* (preferably of roughly equal status) *with quite different skills and backgrounds.* It is important, of course, to be assured that a genuine cooperation is proposed (in fact, is necessary if the stated objectives are to be realized) and that no participant is 'cosmetic'. If the cooperation does not look genuine, this means that the *objective* is not interdisciplinary in its essential nature. To pick an imaginary example out of the air, if someone proposes yet another program on stereology and computer simulation applied to grain growth (an overpopulated field!), it can probably only be justified if a recrystallization expert, a metallographic instrumental expert, and a mathematician/programmer join forces.

- There is a perpetual tension between people who wish to improve an aspect of a sophisticated characterization technique (say, STEM) and those whose concern is to use it. There is much to be said for favoring *proposals, which genuinely combine a user concern with an instrumental innovator.* It is worth remembering that many of our principal techniques arose out of a user's imperative: for instance, Castaing in France developed the electron microprobe analyzer in the late 1940s only because his adviser, André Guinier, wanted to know the composition of GP zones in Al alloys.

- *Look at the past publishing habits of proposers.* If the different components of a nominally interdisciplinary research project by a team of investigators are habitually published by single authors, or maybe by pairs, in a range of quite different journals, that might be cause for concern. If some *at least of the publications are in broad-spectrum journals* (there is no lack of these) *by substantial groups of authors with different forms of expertise, that implies a genuinely interdisciplinary outlook.* Maybe the habit which is beginning to be seen, of specifying authors was equal, then the agonizing about whose name comes first may be less acute! Maybe there could even be an editorial practice of using italics or a different font for the names of authors to whom that consideration applies.

****

Steward Flaschen

Former V.P. for Research, ITT Corporation
Chairman Emeritus, Transwitch Corporation

Interdisciplinary Research–Five Ways to Insure Failure

1. A strong organizational chain of command (only works for the military)
2. Decision by consensus (leads to linear thinking)
3. Listening to the "Marketplace" ("too soon, too soon, soon late")
4. Monthly progress reports (work for two weeks, write reports for two weeks)
5. Annual budgeting (inventing by time table)

****

J. Scott Hauger

Director, Science & Engineering Policy and Practice Group
American Association for the Advancement of Science

In 1996, the American Association for the Advancement of Science (AAAS) established a Research Competitiveness Program under the auspices of the National Science Foundation's Experimental Program to Stimulate Competitive Research (NSF EPSCoR). As part of this program, AAAS staff has visited more than 60 universities in nineteen states, working with junior and senior faculty and research administrators to enhance research competitiveness. We have also conducted five national and regional conferences that addressed issues of collaborative research. These workshops brought together potential research collaborators from different disciplines and different institutions to explore opportunities for interdisciplinary collaboration. Several sessions were devoted to collaboration in areas of materials research. In addition, a science policy conference held in Coeur d'Alene, Idaho, in October 1999, addressed inter-institutional collaboration as one of several strategies for enhancing research competitiveness.

AAAS has reaped a variety of insights into the opportunities and problems of collaborative research as a result of these activities. Several trends are at work in the research community that favor collaboration: First is the increasing breadth and complexity of the scientific community. More specialties in more departments mean that few universities can provide all of the expertise needed to address complex problems. Second is the fact that research problems are becoming more complex. Knowledge accumulation is progressive in the sense that simpler problems are typically solved first, and, over time, a robust science addresses its practices to increasingly sophisticated problems. Third is a modest shift in problem setting from disciplinary needs to issues that are set by social and economic needs identified by communities external to the research institution. The application of the Government Performance and Results Act (GPRA) to institutions funding basic research is a symbol of this third trend of the time. At the same time, the research community has come increasingly to realize that the distinction between basic and applied research is largely a matter of convention. As a consequence, external criteria for problem selection have been admitted to earlier phases of the R&D cycle, increasing the demand for multidisciplinary or interdisciplinary approaches.

Whatever the causes, there is little doubt that research collaboration has been on the rise for decades, a fact that is attested one the one hand, by bibliometric studies, and on the other by the creation of new programs to administer interdisciplinary research, such as the Science and Technology Centers Program, established in 1987, at NSF. At the same time, NSF's Directorate for Computer and Information Science and Engineering (CISE), together with other science agencies, is continuously developing new tools and applications to support research collaboration at a distance. A case could be made that collaboration has become a hallmark of leading edge research in many fields.

AAAS's experience through its collaboration workshops has been that there are many scientists and engineers on college faculties who are eager to undertake collaborative research. There is especially high interest among junior faculty members seeking to establish and build a research portfolio. However, we have also learned that there is often inadequate institutional support to researchers wishing to pursue interdisciplinary collaboration.

At our workshops, many would-be collaborators tell us that their departments discourage interdisciplinary collaboration because they believe that such collaboration hinders (positive) tenure decisions. Tenure decisions are often made by primarily disciplinary committees which tend to discriminate in favor of disciplinary criteria. In addition, some departments maintain that collaboration may represent a crutch, and that it is impossible to judge an individual's contribution within a collaboration.

Even when departments recognize the value and desirability of collaborative research, there are subtle, historically-based barriers to collaboration. Most researchers were trained within a discipline and their knowledge and professional contacts tend to be disciplinary. Many of those attending the AAAS collaboration workshops commented that the workshops represented a unique opportunity to make professional contacts with researchers outside their field who were similarly interested in identifying potential collaborators.

Ironically, it may generally be the case that older, more established researchers have the (tenured) position, the reputation, and the breadth of professional contacts which allow them to undertake the risk of exploring and initiating collaborative research. On the other hand, younger researchers may more often have the motivation, the state-of-the-art knowledge, professional flexibility, and the computer applications skills to comprise the best collaborations. At the beginning of the twenty-first century, the advancement of science and the pursuit of research competitiveness require that department heads, research administrators, and science and engineering policy experts consider the value of interdisciplinary research collaboration and, assuming that value to be significantly positive, work to reform or replace institutional barriers to its practice.

****

W. Lance Haworth
Division of Materials Research
National Science Foundation

We need to bring home the message that materials research generates wealth— a public wealth that is in the national interest—Neal Lane
It is always possible that some enterprising author will find the next big thing hidden away in some scientific backwater, like materials science—John Horgan

The National Science Foundation supports interdisciplinary research and education in materials through a variety of programs in several directorates and divisions. NSF's Division of Materials Research (DMR) was established n 1972 when the Interdisciplinary Research Laboratories were transferred from DARPA and the MRL program was begun, and DMR remains a major focus for materials research and education at NSF. Today, in addition to individual investigator awards, DMR provides explicit support for interdisciplinary research in the field through Focused Research Groups, Materials Research Science and Engineering Centers (MRSECs), and several Science and Technology Centers in materials and closely related areas. There are exciting opportunities in emerging interdisciplinary areas such as nanoscience and engineering, computational materials research, and the interface between materials and biology. DMR also supports user facilities for synchrotron radiation, neutron scattering, and high magnetic fields. Beyond DMR there is also significant support for materials research at NSF through the Chemistry Division, the Engineering Directorate (for example through the Engineering Research Centers) and several other programs. In materials as well as many other areas, NSF is a catalyst for progress and partnerships are crucial for success. Our partners include grantees in colleges and universities of course, and also the private sector, other Federal agencies, State and local agencies and organizations, schools and communities, international organizations, policy makers, and the communications media. Specific examples are given of interdisciplinary research and education supported through various NSF programs including MRSECs, the GOALI program (Grant Opportunities for Academic Liaison with Industry), and partnerships with other organizations.

****

Michael Heylin

Editor-at-Large, Chemical & Engineering News

Institutionalizing Interdisciplinarity
Reprinted from Chemical & Engineering News, Volume 77, Number 37, p. 41

When I see a phrase like the above headline, my eyes tend to glaze over. Apart from the bombastic language, the concept of institutionalizing an approach to doing science designed to lower the barriers between existing institutions has inner contradictions.

Be that as it may, institutionalizing interdisciplinarity was the title of one of the panel discussions at a conference hosted late August by Penn State's Materials Research Laboratory. The gathering, entitled "Interdisciplinarity Revisited," was funded by the National Science Foundation (NSF). It attracted about 90 high-level movers and shakers of interdisciplinary research from around the world for two days of intense and lively debate. The main focus was materials research, a trendsetter for the interdisciplinary approach on campus.

As the conference brought out, interdisciplinarity is a very complex and rather confusing subject. For example:

- Is interdisciplinarity a way for the science community to tackle truly critical national projects and challenges in a timely way?

- Is it just the natural order of things for the manner in which much science is conducted, especially in industry? These days, few if any scientific developments with major impact on society go from concept and fundamental research finding to final application or product all within one discipline.
- Is it a way for government agencies, through their funding mechanisms, to get universities involved in the goal-oriented R&D the agencies are interested in having done?
- Is interdisciplinarity, in the long term, a mechanism to reorder the way academic science is organized and conducted? Is it the way by which new disciplines evolve and older ones get subsumed or fall by the wayside?
- Or is it a way to change the nature of scientific careers? One speaker at the conference proclaimed that the single-discipline scientist is an endangered species.

The short answer is that interdisciplinarity is certainly the first three of the above. The extent to which it is should be, or will ever be the fourth and fifth is contentious.

The archetypal example of the use of the interdisciplinary approach to tackle a grand national challenge has to be the Manhattan Project that produced the atomic bomb during World War II. It was extraordinarily successful scientifically. It advanced from theory to a deployable weapon of a totally revolutionary kind within about three years. And it was all done without computers. Nowadays, it takes a decade or more to develop and deploy a military helicopter.

Ironically, the interdisciplinary Manhattan Project quickly evolved into one of the most secretive and powerful institutions in the U.S., the nuclear weapons community. After 50 years, it still is. This happened despite the initial resistance of many Manhattan Project scientists, who immediately after the war advocated having nuclear research and technology placed under international control, kept as much as possible in the open, and focused on peaceful applications. That was not to be.

There is no doubt that the interdisciplinary approach is the way things get done scientifically in industry. One speaker from a world-renowned industrial R&D operation told conferees of the changes that had taken place over the past decade or so at his laboratory; the shift has been total, from scientists doing truly fundamental research with complete detachment form the company's products or markets to a system of interdisciplinary teams working on product-oriented projects within time frames as short as one year. He claimed it is exciting work and some good science still gets done.

The federal government's approach to interdisciplinary research has been varied. For instance, the Defense Advanced Research Projects Agency invested very early and has been a consistent advocate and funder.

The attitude toward interdisciplinarity at NSF has varied over the years from the extremely chilly to the unenthusiastic and the reluctant. Today, however, although the bulk of NSF funding still goes to scientists in traditional disciplines, the agency is supporting considerable work in interdisciplinary areas, especially materials.

The struggle on campuses over interdisciplinary research is centered on how it fits into, or challenges, the classic academic structure based on disciplines. Disciplines have their advantages. They have traditionally handled academia's primary function of education. They are the keepers of academic values. And they have tight quality control.

The disadvantages of disciplines are that they tend to become vertical and narrow, inner-looking, and reluctant to change. To their critics, they have become self-centered, isolated stovepipes or silos to the detriment of the interdisciplinary approach science needs in today's rapidly evolving technological world. According to one attendee at the conference, tensions have been heightened by hardball campus politics. As he put it, "Only two groups, academia and the Mafia, murder their own."

Interdisciplinary research on campus also has its advantages and disadvantages. According to its advocates, it is flexible, it can handle specific issues of societal concerns quite readily, and it has a track record of achievement despite policies by tenure-awarding departments in the traditional disciplines that discourage faculty participation in it.

On the other side, a persistent concern both on campus and at the funding agencies about interdisciplinary research is the intellectual caliber of those involved and the quality of their work.

The bottom line is that interdisciplinary research is here to stay, on campus and everywhere else. Academia has not passed up the funding offered and over the years has accepted what one conferee estimated as a $1 billion bribe to do it.

Today, there are myriad interdisciplinary projects in academia under various guises. The smart thing would seem to be to have strong departments, strong interdisciplinary activities, and a way for both to benefit their overall institution by dwelling on each others' strengths.

****

Rustum Roy
Evan Pugh Professor Emeritus of the Solid State
The Pennsylvania State University

Scientists to study and evaluate the processes of science management

My first recommendation is devoted to urging individuals, departments, University administrators, *all* Agency personnel to spend 2-5% of their time, studying, discussing, reflecting and writing about the *processes* in which they are involved daily and deeply in doing their science. Some examples are: defining disciplines, departments, interdisciplinary, multidisciplinary, peers, examining major funding policies. Are present allocations among fields justifiable? Do individual investigators work to group work, such as: evaluating funding mechanisms after reading the literature; formula funding, institutional funding, strong managers (as in DoD); and mail peer review, etc. policies governing university-industry interactions, intellectual property regulations on local campus, and options elsewhere. A great lacuna in the body politic of science is the near total non-participation by the vast majority of scientists in their own governance, and this recommendation aims to remedy that.

Agencies should fund studies of the most successful programs, which have achieved genuinely interactive research

It is clear that no Agency has any databased empirical study on what administrative features are correlated with *sustainable* patterns of interdisciplinary or more precisely, I³R, on a campus. This is a simple matter; methodologies exist for ranking everything; it is an extraordinary omission. In fact a NATO group did precisely this in the seventies. In the eighties when I chaired the National Academies' U.S.–USSR exchange program committee, we had U.S. scientists write an evaluation of their host institution's "quality" after their stays. The President of NAS wished these to be held confidential, but the Foreign Secretary of the Soviet Academy begged us to release them to their office, because they knew it would be the most objective evaluation possible, by knowledgeable scientists, with no possible conflict of interest.

Soon thereafter when requested by the Swedish Academy, I recommended exactly the same objective *international team visits*, as the best way to evaluate their chemistry departments. After the process the Academy President reported that those evaluated felt that the process was very good.

Create administrative structures before writing proposals

Universities should be *required* to establish their own appropriate interdisciplinary *structures* for their local situations in specific fields, BEFORE applying for grants in the field. The particular University's commitment to interdisciplinary work would be then evident to the reviewer, rather than being camouflaged in a series of maybes and possibilities and ONLY with SOFT MONEY. Likewise for University-Industry work, the intellectual property posture should be made explicit and agreed to by the PI and the corporate contacts.

Provide *evidence* that the University has leveled the playing field between disciplines and interdisciplinary activities. Before providing funds an Agency should determine: What incentives are given to faculty to run against the tide of disciplinarity? Such as:

- Counting I²R papers double in departmental evaluation
- Matching funds (or overhead forgiveness) for collaborative work with industry or government
- Travel funds to defray cost of collaboration with expensive facilities available at other Universities (much more cost effective than maintaining white elephants used 5 hours/week)
- Tenured and chair professorships explicitly located in locally selected interdisciplinary fields
- Graduate assistantships/fellowships available for emphasizing specific I.D. fields
- University should provide evidence of existing (not future) incentives to broaden faculty's interest horizon. E.g. rewarding I³R work of any kind requiring some teaching, research or public service outside of department for Distinguished Ranks, and for future emeritus status.

****

Helmut K. Schmidt

Institut fuer Neue Materialien
Saarbruecken, Germany

Materials play an important role in industrial innovation. But materials are only an intermediate step to components and systems. From the basic research to the market place, many disciplines are involved. Following the added value chain through the different steps to the final product, one can easily see that the material is positioned at the first (lowest) step. This leads to a difficult situation for the material developer: The development of the material takes long time periods, connected with high risks and high costs. This means that only those materials are industrially developed which promise high market volumes. On the other hand, basic research at universities and national R+D centers produces many new and interesting materials. To exploit these developments, facilities are required which permit to develop appropriate materials and application technologies at the users' place (since producers only step in for large markets). These developments, however, in general, can be carried out at very large system manufacturers like electronic, aviation, space or automotive industries. Summarizing, one can say that the technology development gap (TDG) between new material basic research and potential users is one of the most important drawbacks in new materials industrial exploitation.

In order to overcome the TDG problem, a concept for a public R+D center has been developed to combine basic research with technology development at the same place. The concept comprises the personnel aspect (interdisciplinarity), the marketing aspect (market searches, connection to industries), the management aspect (organization rules, form of the R+D facility), the technical and scientific aspect (choice of appropriate themes, technical equipment), the strategy (e.g. balance of basic and contract research, chemical materials synthesis as a basis) and contract and license philosophy. Based on these considerations, an organization in the form of an inc. ltd. has been founded. As technology basis, the chemical synthesis of nanomaterials has been chosen. As the most promising areas for industrial applications, wet chemistry coating technologies were identified by market searches. As material basis, glasses, ceramics and nanocomposites seemed to be of interest. After a period of three years of basic research, collaborations with industry have been started. At present, more than 100 projects are carried out at the same time by a staff of over 250 people. Meanwhile, turn-key technologies can be developed as well as it is possible to provide appropriate interfaces to fulfill all requirements from industry. For the material production, strategic partners have been acquired to guarantee the materials supply outside of mass commodities.

Examples for successful developments are the development of an optical sealing system for fiber-to-chip coupling, based on a nanocomposite, a transparent hard coating for PMMA (lenses), PC and CR39, an enamel substitution on metals by a nanocomposite hard coating, an SiC ignitor for gas firing, a deodorizing catalyst for kitchen equipment and an easy-to-clean nanocomposite coating for sanitary and bathroom applications. In the described examples the scientific and technology challenge, the solutions, the strategic approach and the route for a successful realization will be presented.

****

Yu. D. Tretyakov

Founder, Degree Program in Materials Science
Moscow State Lomonossov University

- In addition to three dimensions of I³R model it seems to be reasonable to add the fourth dimension—*active Inter-national Materials Research cooperation among universities.*
- *Only due to such Inter-national Materials Research cooperation the world community would be able to obtain a profit from recent situation in*

Russia whose rich creative potential used now ineffectively inside the country because of industry decline and permanent political crisis.

- It makes much more sense to support the active part of the Russian research and education system because it acts as a kind of counterweight to the mafia, which is expanding its influence inside and outside the country.

- It would be reasonable to convert the negative stochastic "brain drain" process in mutually profitable one *by training Russian materials researchers in accordance with US industrial or academic demands.*

- The satisfaction of these demands could be compensated by support of Russian interdisciplinary institutions like Higher School of Materials Science.

- It would be *reasonable to develop international materials research cooperation* in terms of NSF, ISF, CRDF, NATO, INTAS projects with active Russian participation.

****

Hiroaki Yanagida

Former Director, RCAST, University of Tokyo
Professor Emeritus, University of Tokyo

The symposium was very interesting not only discussion about scientific facts but also philosophy and strategy. *I learned that interdisciplinary researches are usually more effective than mono-disciplinary research at least to find novel phenomena or functions. However, interdisciplinary ones have always to struggle to survive. Otherwise they will disappear.* Ordinary scholars and researchers are comfortable when they stay within their own discipline. And sometimes they are the opponents against new disciplines, which may lessen the value of their own discipline. Even students stick to, or are inclined to define themselves as belonging to the fields, which are already well established.

Disadvantages are observed in researches driven by mono-disciplinary champions especially in cases *where the work is needed by society or the general public*. Things well accepted in one discipline may not be accepted by other disciplines. Debates must be raised between different disciplines. Interdisciplinary work is not limited to the sciences and technologies. *I propose that we one must take into consideration* the different *contributions by professionals and amateurs, social sciences and natural sciences, sciences and technologie*s, classical or traditional wisdom and advanced technology, academia and industry, etc.

As far as materials sciences are concerned, interfaces between different materials including different crystal orientations can be very fruitful sources of novel findings. The frontier ceramic project sponsored by Science and Technology Agency of Japan Government (led by Prof. Yanagida) is one of the typical examples. This is extended to a new science or technology area.

****

Robert E. Yager

Professor of Education, University of Iowa
President of the National Association of Science, Technology and Society

Recommendations for Policy Maker
- Look for *ways (monetary and recognition) that would encourage moves on college campuses to interdisciplinary instructional* programs and research
- Encourage more research that will provide needed *evidence of the values to students of pursuing interdisciplinary studies*
- Promote publications that are collections of "stories" on approaches and results of interdisciplinary experiences for teachers and students

Ideas for Overcoming Departmentalism

Departments in a traditional sense may have outlived their usefulness. They may make sense for administrative purposes, e.g. devising schedules, defining courses, managing budgets, encouraging communication. However, *rarely are departments seen as units for initiating action*, focusing on projects, and establishing better contexts for learning.

Departments may be better conceived of, as existing for given purposes, e.g., to deal with given projects, to meet student goals, to undertake specific tasks. They could draw people with varying talents and expertise to meet certain goals.

Charter schools in the education arena provide examples. They are schools drawing teachers because of their interest in new approaches to realizing state goals. These are often goals that can not be met in the typical organizational system of a school where there is a tenure staff and administration that are not in tune with the new goals and/or projects.

A second idea for overcoming departmentalization is to recognize the importance of context in stimulating and achieving new learning. Situations can be established to illustrate how this focus could overcome barriers.

The third idea for overcoming the perils of departmentalization is recognizing the importance of technology as a connector between science and the whole of society. This will not be an issue in a College of Engineering. However, in typical high schools and undergraduate programs in arts and science, technology never gets considered other than educational technology (the compute). Nonetheless, for most people the value and importance of technology far surpasses the value and importance of science per se. In fact, research shows that more people (high school and college students) are more knowledgeable and positive about technology than they are of science—even though most have not studied technology in the formal sense.

****

Appendix I

Affiliations of Participant Authors

Affiliations of Participant Authors of Papers ▶▶

Arden Bement, Department Chair Nuclear Engineering, Purdue University; Former Chair NRC Committee on Materials Innovation

Richard J. Brook, Chief Executive, Engineering and Physical Sciences Research Council, London, England

George Bugliarello, Chancellor and Former President, Polytechnic University of New York; Founding Editor of *Technology and Society*; Editor of National Academy of Engineering's "Bridge"

Robert Cahn, Department of Materials Science & Metallurgy, University of Cambridge, England; Founding Editor of *Journal of Materials Science*; First Head of Materials Science Department, Sussex University

K.K. Deb, Army Research Laboratory, Sensors and Electron Devices Directorate

T. Egami, Professor of Materials Science and Engineering, University of Pennsylvania

Rodney A. Erickson, Provost, The Pennsylvania State University

Irwin Feller, Director, Institute for Policy Research & Evaluation, The Pennsylvania State University

Steward S. Flaschen, Former V.P. for Research, ITT Corporation; Chairman Emeritus, Transwitch Corporation

Tracy Gaudet, Co-Director, Program in Integrative Medicine, University of Arizona

Scott Hauger, Director, Research Competitiveness Program, AAAS

Lance Haworth, Executive Officer, Division of Materials Research, National Science Foundation

Michael Heylin, Editor-at-Large, *Chemical & Engineering News*

Joyce Jentoft, Associate Provost and Dean of Graduate Studies, Case Western Reserve University

Mohammad Karim, Chair of the Department of Electrical and Computer Engineering, University of Tennessee

Elton Kaufmann, Associate Director of Argonne National Laboratories; Founding Editor, *Materials Research Society Bulletin*

Shigeyuki Kimura, Director, National Institute of Research on Inorganic Materials, Tsukuba, Japan

Amitabha Kumar, V.P for Product Engineering, E. Khashoggi Industries, Santa Barbara, California

Robert McGrath, Assistant Vice President for Research, The Pennsylvania State University

Arumugam Manthiram, Texas Materials Institute, The University of Texas

Katy Marre, Associate Vice President for Graduate Studies and Research, University of Dayton

Paul C. Maxwell, Vice President for Research and Sponsored Projects, University of Texas at El Paso

Larry Murr, Dept of Metallurgical and Materials Engineering, Materials Research Institute, The University of Texas at El Paso

P.S. Nicholson, Professor, Department of Materials Science and Engineering, McMaster University

Forrest J. Remick, Former Commissioner, Nuclear Regulatory Commission; Former Associate VP Research, The Pennsylvania State University

Rustum Roy, Evan Pugh Professor of the Solid State Emeritus and Professor of Science, Technology and Society Emeritus, The Pennsylvania State University; Founding Director, Materials Research Laboratory; Founding Editor, *Materials Research Bulletin* **and**

Materials Research Innovations; Principal Architect of the Materials Research Society

Manfred Rühle, Director Max-Planck-Institut für Metallforschung, Stuttgart, Germany

Helmut K. Schmidt, Professor, INM Institut für Neue Materialien, Saarbrücken, Germany

Lyle Schwartz, Director, Aerospace and Materials Sciences, AFOSR/NA, Former Head, Materials, N.I.S.T.

Thomas Stoebe, Former Department Chair of Materials Science and Engineering, University of Washington

Yuri Tretyakov, Professor of Chemistry and Founder, Degree Program in Materials Science, Moscow State Lomonossov University

Andrew T. Weil, Program in Integrative Medicine, University of Arizona

Robert Yager, Professor of Education, University of Iowa; President, *National Association for Science, Technology and Society*

Hiroaki Yanagida, Former Director, Research Center for Applied Science and Technology, University of Tokyo; Professor Emeritus, University of Tokyo

Appendix II

Case Studies of I³R in Materials

Interdisciplinarity in the Development of High Energy Density Batteries ▶▶

Arumugam Manthiram
Texas Materials Institute
The University of Texas at Austin

The exponential growth in portable electronic devices such as cellular phones and laptop computers and the drive to develop electric vehicles have created enormous interest in high energy density batteries. Lithium-ion batteries are attractive in this regard, as they offer higher energy density compared to other rechargeable systems, and are emerging the choice to power portable electronic devices. The success in lithium-ion battery technology is a result of true interdisciplinary interaction for three decades among different disciplines. Innovative materials design based on fundamental physics and chemistry concepts, development of novel synthesis and processing procedures, advanced materials characterization, fundamental understanding of the solid state electrochemical phenomena, engineering design and fabrication of electrochemical power sources (batteries) have all played a key role in the success.

The currently available commercial lithium-ion cells utilize a lithium cobalt oxide cathode, but cobalt is expensive and relatively toxic. Also, only 50% of the theoretically available capacity of the cobalt oxide could be practically utilized due to the chemical and structural instabilities encountered at deep charge exceeding the extraction of 0.5 lithium per cobalt. In view of the rapidly growing consumer market in portable electronics and large-scale batteries for electric vehicles, the future challenge is in the development of inexpensive and environmentally benign electrode hosts of high capacity and energy density. Manganese and iron oxide cathodes are attractive in this regard, but the known manganese and iron oxides do not offer satisfactory performance. In the case of manganese, the intensively pursued lithium manganese oxide spinel tends to exhibit capacity fade during cycling due to lattice distortion and manganese dissolution.

We have developed an amorphous manganese oxide cathode employing a novel solution-based synthesis approach in nonaqueous medium. The amorphous manganese oxide exhibits excellent electrochemical cyclability with a high capacity that is nearly two times higher than that of the currently used lithium cobalt oxide or the intensively pursued lithium manganese oxide spinel [1, 2]. However, the amorphous manganese oxide exhibits a sloping voltage, which may be a drawback. We have also developed a crystalline manganese oxide spinel with a higher oxidation state of 4+ for manganese by another novel synthesis approach [3]. This spinel oxide also exhibits excellent electrochemical cyclability with a capacity that is 30% higher than that of the conventional lithium manganese spinel oxide, but with a decrease in the cell voltage by 1 V.

Our research in lithium-ion batteries involves the complete cycle of materials design, synthesis, characterization and performance evaluation. For example, the research involves design based on chemistry and physics concepts, innovative chemical synthesis to access the designed material, advanced chemical, structural and thermal characterizations, electrode fabrication and battery construction, electrochemical and transport measurements, and an understanding of the structure-property-performance relationships. It is difficult to carry out this interdisciplinary research activity and achieve the goals with graduate students having prior degrees in one of the traditional disciplines. Also, as a faculty member in the Department of Mechanical Engineering, it is even more difficult in my case to get graduate students who will be able to perform this interdisciplinary research.

The barrier was, however, overcome by drawing graduate students through the interdisciplinary Materials Science and Engineering Graduate Program, which is not a Department at the University of Texas at Austin. Materials Science and Engineering Graduate Program enabled to have students with prior degrees in different disciplines working in my laboratory. Access to students with different background and training them in a broader interdisciplinary activity have contributed to our success in developing low cost and environmentally benign cathodes of high capacity and energy density for lithium-ion batteries.

References:

1. J. Kim and A. Manthiram, Nature 390, 265 (1997).
2. J. Kim and A. Manthiram, Electrochem. Solid State Lett. 2, 55 (1999).
3. J. Kim and A. Manthiram, J. Electrochem. Soc. 145, L53 (1998).

Conductive Protein Films for Uncooled IR Microbolometer Arrays ▶▶

K. K. Deb
Army Research Laboratory
Sensors and Electron Devices Directorate

Abstract

The detector sensitivity of an infrared (IR) sensor, which is based on a silicon microbridge microbolometer, is currently limited by the low temperature coefficient of resistance (TCR) of the active element. Large research efforts have been devoted to the development of new bolometer materials with larger TCRs to the ambient region for improving the present uncooled technology to be competitive with the cooled detectors. I demonstrate that cytochrome c and bovine serum albumen monolayers on silicon substrate create new materials for high performance IR bolometers because of the higher TCR values associated with these proteins.

Measurements were performed on a 40 A thick cytochrome c film deposited on a Si/SiO_2 structure. The film is quite predictable and exhibits very high TCR at ambient temperature. An observed value of the room temperature TCR is an order of magnitude higher than the present technology. The film exhibits a positive TCR between ambient temperature and 90 ^{0}C. The TCR value depends on the forcing current with higher values measured for forcing currents up to 100 mA. With 1 mA current, the sheet resistance of the film is about 1 ohm/cm^2. Proteins are self-absorbing between the 3- to 5- mm and 8- to 12- mm regions and no additional IR absorber coating is required to achieve efficient radiation capture. All these advantages will make the protein IR microbolometer technology very promising for future military and commercial sensor applications.

How Neutron Scattering is Helping to Clean up the Air ▶▶

T. Egami*, R. Brezny, R.J. Gorte, and J.M. Vohs
*Department of Materials Science and Engineering
University of Pennsylvania, Philadelphia

A tool of basic science, neutron scattering, has proven to be quite effective in addressing the most practical question of how to improve automotive catalytic converters that clean up engine exhaust emissions. This is done as a part of an interdisciplinary effort bridging materials physics, chemical engineering, and chemical and automotive industries. Automotive exhaust emissions include various toxic gasses, such as CO and hydrocarbons that have to be oxidized, and NO_x that has to be reduced. This acrobatic feat of simultaneous oxidation and reduction can be achieved by fine particles of transition metal catalysts, such as Pd/Rh, but only when the oxygen partial pressure is within a very narrow window. In order to keep the supply of oxygen within a fraction of a second, ceria, CeO_2, is being used as a catalytic support. However, special processing such as nano-crystalline formation and mixing with zirconia (ZrO_2) is required for the ceria to be effective. Neutron scattering experiment using pulsed neutrons coupled with the real-space analysis using the atomic pair-density function (PDF) analysis has been quite effective in revealing the microstructural features necessary for ceria to act effectively as oxygen buffer. An industry even purchases neutron beamtime from a national laboratory to carry out proprietary works of characterization. We present some of these results and discuss how basic tools can be effective for understanding advanced materials used in various technologies today.

A New Foam Material: A Case Study of Interactive Research in Materials Science and Engineering ▶▶

Per Just Andersen, Amitabha Kumar*,
and Simon K. Hodson
E. Khashoggi Industries
Santa Barbara, CA

The development and mass manufacture of a new foam material with specific application as packaging material is discussed. The primary goal is to mass produce disposable packaging materials from renewable resources that are low in cost, low in environmental impact, and meet commercial performance requirements. Motivation for research is attributed to market pull, technology push and entrepreneurial initiative. Selection of research areas and process adaptations are driven by concerns for conservation of the global environment. The need to tailor materials and processes to meet market expectations is discussed in this context.

The case study describes initial development strategies that include experimentation in chemically bonded ceramics and analogous systems. Product development through interdisciplinary materials research is described. Innovative engineering developments required for successful manufacturing processes, product quality, mass production and consumer satisfaction is detailed.

Environmental life-cycle models and market driven economic models are additional tools required for successful research, development and commercialization. Mass production of the resultant highly inorganically filled starch based fiber reinforced foam composite material designed for food-packaging use is described.

Music, Physics, and Materials Research: The Caribbean Steel Drum as an Interdisciplinarity Model ▶▶

L.E. Murr and E. Ferreyra
Department of Metallurgical and Materials Engineering
Materials Research Institute
The University of Texas at El Paso

Early rhythm metal drums made from baking tins and paint cans with bulges on their surfaces to create specific pitches evolved into larger platforms utilizing 55 gallon (steel) oil drums which provided primary fuel storage on Caribbean islands such as Trinidad and Tobago around 1945. Over the past 50 years, these musical steel drums have evolved into essentially nine orchestral voices, with drum note patterns ranging from 3 to 32 notes; covering a tonal range of A1 to F6 (55 Hz to 1397 Hz). In 1996, an interdisciplinary team of musicians, physicists, and materials scientists and engineers was formed at UTEP to develop a materials research program to elucidate the structure, properties, and performance issues unique to the fabrication, processing, tuning, and acoustic qualities of the Caribbean steel drum. Artistic performance was also combined with materials performance analysis to create a convergence of art and science under the banner of interdisciplinary materials research. TEM analysis of extracted musical notes coupled with static and dynamic microhardness mappings and light metallography of complete drum and note surfaces provide unique insights into the plastic-elastic interactions which contribute to non-linear acoustic vibrational phenomena characteristic of the Caribbean steel drum. The effect of heating the patterned steel drum head over a fire-pit or similar arrangement has been determined to result by strain aging, which is optimized by some requisite carbon content in the steel. This contributes to enhanced tuning and musical qualities and is a crucial issue in producing superb sounds. Supported in part by a Murchison Endowed Chair and a Shell Oil Company Foundation Grant.

Thermoluminescent Dosimetry: A Joint Develoment of Materials Science and Health Physics ▶▶

Tom Stoebe
Materials Science and Engineering
University of Washington

Materials science has a reach that is both broad and deep. In the area of radiation dosimetry, the common method today is the use of thermoluminescence, with the most common luminescent materials being LiF. LiF is an alkali halide which 30 years ago, was studied as an idea material. It's mechanical, optical and defect properties were studied by physicists and materials scientists, but no practical use was envisaged.

However, physicists at the University of Wisconsin, first, led by John Cameron, found that the luminescent property of LiF contained several advantages that could be used for personnel radiation dosimetry, namely: that it has a effective Z number nearly identical to that of water and therefore of humans,

- that it has a effective Z number nearly identical to that of water and therefore of humans,
- that with certain impurities it stored radiation information as luminescence and the luminescence output was proportional to the prior gamma radiation dose,
- that this information was stable and could be stored for many years without degradation, and

• that this information could be detected easily by heating, that is, by measuring the thermoluminescence output of the material.

This led to the development of today TLDs, which are used universally in personnel radiation detection today.

However, this could not have happened without the understanding of this material from the point of view of its defect properties. Health physicists struggled in the early days of TLDs trying to understand changes in properties found after different annealing and cooling treatments, for example, and many made the simple error of considering that radiation history, rather than changes in defect structure, were the cause of the variations seen.

It was only with the application of materials science principles that TLD became trusted as the best and most reliable way to measure gamma radiation dose. This work by no more than a handful of materials scientists, allowed for understanding of heat treatment variables, and thus of changes in defect structure that were controlled by the heat treatments and eventually controlled the sensitivity of this material. A similar story can be said for the understanding of the impurities and impurity concentrations that provide for maximum use fullness.

This paper will follow the history of materials science innovations that allow use of TLDs by health physicists today.

Appendix III

Final Program: Interdisciplinarity Revisited Conference

Final Program

International Conference on

Interdisciplinarity Revisited

Materials Research as a Case Study

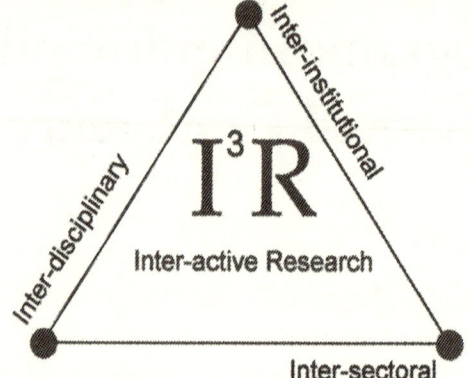

August 30-31, 1999

Penn Stater Conference Center Hotel
The Pennsylvania State University

Sponsored by
National Association of Science, Technology and Society

Financially supported by
The Alfred P. Sloan Foundation and the National Science Foundation

Materials Research
Laboratory

Conference Venue & Host

The Penn Stater Hotel and Conference Center:

Perched on a knoll overlooking Mt. Nittany, in University Park, which is part of the borough of State College, Pa., our conference venue is ringed with jogging and biking trails. Inside you may take a few laps in the pool, unwind in the whirlpool, or relax in the sauna. A fully equipped fitness center can take the edge off a long day and keep your performance high. The Gardens Restaurant serves international cuisine, with an emphasis on seafood. Hours: Breakfast, Monday-Friday 6:45-11:00, Sunday 6:45-10:00, Brunch 10:00- 2:00; Weekly Lunch 11:30-2:00; and Dinner nightly 5:30-9:00. Legends Pub is for relaxing, lighter fare, and a friendly game of darts or billiards. Hours: Sunday-Saturday, 11:30-11:00 p.m.

The Materials Research Laboratory:

The host for this conference is Penn State's Materials Research Laboratory (MRL). Located on the east side of the campus of The Pennsylvania State University, the Laboratory was founded in 1962, the first such established by a U.S. University, independent of any grant, on the premise that collaboration across academic disciplines and with industry could accelerate the pace of materials development and implementation. It has been among the largest and most influential MRLs in the U.S. The Monday dinner event will be al fresco, outside the Materials Research Laboratory, with an opportunity to visit the laboratory, immediately after.

Quick Summary of Conference Schedule

Registration:

Time	Place
Sunday 10:00 a.m.-1:00 p.m.	The Materials Research Laboratory
Monday 7:00 a.m. – noon	Penn Stater Hotel and Conference Cnt., Lobby in Front of Session Room

Conference Events:

Time	Place
Sunday 10:00 a.m. – 10:00 p.m.	**Pre-Conference Events (SEE SEPARATE SHEET)**
Monday 7:00 – 8:30 a.m.	**ALL SESSIONS (excluding Monday dinner) in Conference Center** **Registration**, Located in Foyer outside of *Conference Room G* Posters and Continental Breakfast in *Conference Room F*
8:30 – 12:30 p.m.	**Session I: Present at the Creation**, *Conference Room G*
12:30 – 1:30 p.m.	Lunch Break, *The Gardens Restaurant*
1:30 – 5:30 p.m.	**Session II: I³R: the Status Worldwide**, *Conference Room G*
6:15 – 8:30 p.m.	Reception, Buffet Dinner *al fresco* and Lab tour at the *Materials Research Laboratory*
Tuesday 8:00 – 12:00 p.m.	**Session III: Outstanding Successes in I³R**, *Conference Room G*
12:00 – 1:00 p.m.	Lunch Break, *The Gardens Restaurant*
1:00 – 4:00 p.m.	**Session IV: Output to R & D Administrators and the World**, *Conference Room G*

Monday
August 30, 1999

REGISTRATION
7:00 a.m. – 8:15 a.m.
Foyer of Plenary Room
at the Penn Stater
Conference Center

Continental Breakfast Provided
Conference Room F

PROGRAM
8:15 a.m.

Welcome

Robert McGrath, *Assistant Vice President for Research, The Pennsylvania State University*

Scope of this Conference

Rustum Roy, *Evan Pugh Professor of the Solid State Emeritus and Professor of Science, Technology and Society Emeritus, The Pennsylvania State University Chair, Interdisciplinarity Revisited Conference*

SESSION I:
PRESENT AT THE CREATION
9:00 a.m. – 12:30 p.m.

Historical Perspectives From Those Present at the Beginning:

W.O. Baker, *Former President, Bell Telephone Laboratories, Advisor to the White House occupants for four decades. Chair NRC COSMAT report*

Frederick Seitz, *Former President, National Academy of Sciences, a founder of the field of Solid State Physics*

Arthur Von Hippel, *(via message read by Professor Harry Gatos), Former Director, Laboratory for Insulation Research, M.I.T. (First Interdisciplinary Unit on a campus.)*

Panel 1: The Driving Forces for/ Barriers to, Interactive Research

Lyle Schwartz, *(Government) Director, Aerospace and Materials Sciences, AFOSR/NA. Former Head, Materials, N.I.S.T.*

Mark Myers, *(Industry) Executive Vice President for Research, Xerox Corporation*

Arden Bement, *Department Chair Nuclear Engineering, Purdue University, Former Chair NRC Committee*

Forrest J. Remick, *Former Commissioner, Nuclear Regulatory Commission, Former Associate VP Research, Penn State*

Break 10:45 a.m. – 11:15 a.m.

Panel 2: Driving Forces for/ Barriers to, Interdisciplinarity in Universities: Teaching and Public Service

Robert Cahn, *Department of Materials Science & Metallurgy, University of Cambridge, England, Founding Editor of J. Materials Science*

Merton Flemings, *Department of Materials Science, M.I.T., Former Department Chair, Co-Chair NRC report*

Eric Baer, *Institute for Macromolecular Science, Case Western University*

Yuri Tretyakov, *Founder, Degree Program in Materials Science, Moscow State Lomonossov University*

Responses by:

Katy Marre, *Associate Vice President for Graduate Studies and Research, University of Dayton*

Mohammad Karim, *Chair of the Department of Electrical and Computer Engineering, University of Tennessee*

Lunch Break 12:30 - 1:30 p.m. The Gardens Restaurant

Monday
August 30, 1999

Yet, as we seek to reconstruct the engineering curriculum, the legacy is a disaster. In the words of a retired philosophy professor: At the core of the problem is a combination of both disciplinary hubris and academic territoriality, both militating against cooperative ventures across disciplinary and administrative boundaries.
— *Samuel Florman, ASEE Prism 1997*

SESSION II:
I³R: The Status Worldwide
1:30 – 5:30 p.m.

Chair

L. **Eric Cross**, *Evan Pugh Professor Emeritus of Electrical Engineering, Chair 1ˢᵗ Annual meeting of MRS*

Overview

Richard J. Brook, *Chief Executive, Engineering and Physical Sciences Research Council, London, England*

Panel 3: The World Status of I³R

P.S. Nicholson, *(Canada) Professor, Department of Materials Science and Engineering, McMaster University*
S. Arunachalam, *(India) Department of Materials Science and Engineering, Carnegie Mellon University, Former Head, Defense Science Establishment, India*
Hiroaki Yanagida, *(Japan) Former Director, RCAST, University of Tokyo, Professor Emeritus, University of Tokyo*
Manfred Rühle, *(Germany) Director Max-Planck-Institut für Metallforschung, Stuttgart, Germany*
Chenzhi Li, *(China) Minister for Education, Peoples Republic of China*

Comments and Discussion from the Floor

Break 3:00 – 3:30 p.m.

Panel 4: I³R: In Other Subject Areas

- **Social Sciences** — **Irwin Feller**, *Director I.P.R.E., The Pennsylvania State University*
- **Integrative Medicine** — **Tracy Gaudet**, *Co-Director, Program in Integrative Medicine, University of Arizona*
- **Science, Technology and Society** — **Robert Yager**, *Professor of Education, University of Iowa, President National Association for Science, Technology and Society*

Comments and Discussion from the Floor

Reception and Buffet Dinner *al fresco*
6:15 p.m. – 7:30 p.m.
Introduction: G.L. Messing, Director, Materials Research Lab

Materials Research Laboratory
Lab Tours following Dinner
Visits to Specific labs or groups, or general tour

(Van transportation available, see conference registration desk)

Tuesday
August 31, 1999

A particularly important development, in my view, was the Open University. This remarkable university concentrates on "distance learning" for mature students who cannot attend a normal university, because they are in fulltime employment, or married with children, or (most commonly) do not have formal school-leaving qualifications. The founding fathers of the OU laid it down that "we take it as axiomatic that no formal academic qualifications would be required for registration, and only failure to progress adequately would be a bar to continuation of studies," and this principle has been adhered to for 30 years.

— R. W. Cahn

SESSION III:
Making I³R Work
8:00 a.m.—12:00 p.m.

Chair

Robert E. Newnham, *ALCOA Professor Emeritus, The Pennsylvania State University*

Overcoming Barriers: The Agents of Change Toward I³R

Panel 1: *Role of the Government (U.S. Agency Reps)*
Robert Pohanka, *Director of the Materials Science Division, Office of Naval Research*
Jane Alexander, *Deputy Director of Defense Advanced Research Projects Agency*
Lance Haworth, *Executive Officer, Division of Materials Research, National Science Foundation*

Panel 2: *Role of Industry*
Alastair Glass, *Director, Photonics Research Labs, Bell Labs, Lucent Technologies*
Shigeyuki Kimura, *Director. NIRIM, Tsukuba. Japan*
H. K. Schmidt, *Professor, INM Institut für Neue Materialien, Saarbrücken, Germany*

Break 9:45 — 10:15 a.m.

Panel 3: *Role of Professional Societies and Their Journals*
The Materials Research Society (MRS, I.U.M.R.S., and the Science Press)
H.C. Gatos, *1ˢᵗ President MRS, former Associate Director of MRL, M.I.T.*
R.P.H. Chang, *Director MRL, Northwestern University, General Secretary of I.U.M.R.S., Former President of MRS*
Elton Kaufmann, *Associate Director of Strategic Planning, Argonne National Laboratories, Editor, Annual Review of Material Science, one of the founders of the MRS Bulletin*
Scott Hauger, *Director, Research Competitiveness Program, AAAS*
Wil Lepkowski, *Former Senior Editor, C&E News*
Renée Ford, *Former Editor of Materials Technology*
Founding Editors *of J. of Materials Science, Materials Science & Engineering, Surface Science, J. of Materials Research Ferroelectrics, Porous Materials, Materials Research Innovations, Materials Technology, Technology & Society, Materials Research Bulletin*

Tuesday
August 31, 1999

> It is by now abundantly evident that interdisciplinary or interactive research in the broad sense used in this conference is here to stay. This is so even if our knowledge institutions are still by and large only timidly approaching the issue of how to encourage interdisciplinarity.
> —George Bugliarello

> In many disciplines outside the hard sciences it is said that we live within "fields" of thought and perception -- nonmaterial, invisible forces that structure both space and behavior within it.
> —Jeff Gates, President, Shared Capitalism Institute

Panel 4: *Lessons Learned from Real-Life Examples of I² Research (5 min. concentrates + posters)*

Larry Murr, Dept of Metallurgical and Materials Engineering, and Materials Research Institute, The University of Texas at El Paso, "Music, Physics, and Materials Research: the Caribbean Steel Drum as an Interdisciplinarity Model"

Amitabha Kumar, V.P for Product Engineering, E. Khashoggi Industries, "A New Foam Material: A Case Study of Interactive Research in Materials Science and Engineering"

T. Egami, Dept. of Materials Science and Engineering, University of Pennsylvania, "How Neutron Scattering is Helping to Clean Up the Air"

Tom Stoebe, Dept. of Materials Science and Engineering, University of Washington, "Thermoluminescent Dosimetry: A Joint Development of Materials Science and Health Physics"

Arumugam Manthiram, Texas Materials Institute, The University of Texas, "Interdisciplinarity in the Development of High Energy Density Batteries"

K.K. Deb, Army Research Laboratory, Sensors and Electron Devices Directorate, "Conductive Protein Films for Uncooled IR Microbolometer Arrays"

Gary Vezzoli, Research Physicist and Instructor of Mathematics, Institute for Basic Science, "The Interdisciplinary Approach to Understanding the Physics of High-Temperature Superconductors"

B.L. Oksengendler, Institute of Chemistry and Physics of Polymers, Uzbekistan, "On Adaptive Material Science"

Lunch Break 12:00 - 1:00 p.m., The Gardens Restaurant

SESSION IV:
Output to R & D Administrators and the World
1:00 - 4:00 p.m.

Chair

Delia M. Roy, *Professor Emerita of Materials Science, The Pennsylvania State University*

- **The Interdisciplinarity Imperative**
  George Bugliarello, *Former President, Polytechnic University of New York, Founding Editor of "Technology and Society, Editor of NAE "Bridge"*

- **Re-inventing Interdisciplinarity**
  C.K.N. Patel, *Vice President for Research, UCLA, (Chair: "Re-inventing the Research University" Conference)*

Tuesday
August 31, 1999

... Materials play an important role in industrial innovation. But materials are only an intermediate step to components and systems. From the basic research to the market place, many disciplines are involved. Following the added value chain through the different steps to the final product, one can easily see that the material is positioned at the first (lowest) step. ... On the other hand, basic research at universities and national R & D centers produces many new and interesting materials... Summarizing, one can say that the technology development gap (TDG) between new material basic research and potential users is one of the most important drawbacks in new materials industrial exploitation.
— Helmut K. Schmidt

The Focus Groups on each of the I^3 Interactions
(Rustum Roy, Interlocutor)
Each group will be asked to respond to the question: What would you recommend to enhance interactive research today, in the light of past difficulties?

Panel 1: *Overcoming Departmentalism: Institutionalizing Interdisciplinarity*

Richard Brook, *Chief Executive, Engineering and Physical Sciences Research Council, London, England*
Robert Yager, *Professor of Education. University of Iowa, President National Association for Science, Technology and Society*
Irwin Feller, *Director I.P.R.E., The Pennsylvania State University*
S. Mahajan, *Department of MSE, Arizona State University*

Panel 2: *Smoothing the Interface: Univ.-Industry; Univ.-Government; Government-Industry Coupling*

J. Mazumder, *Director, Center for Laser-assisted Mfg., University of Michigan*
Kathy Taylor, *Head, Physics & Chemistry Laboratory, General Motors*
Steward S. Flaschen, *Former V.P. for Research, ITT Corporation, Chairman Emeritus, Transwitch Corporation*
Arthur Diness, *Institute for Defense Analysis, Former Deputy Director for Engineering, Office of Naval Research*
Nava Setter, *Director EPFL, Lausanne, Switzerland*

Panel 3: *Overcoming the Bias Toward 'Basic Science': Changing National Attitudes To Applications-Driven Basic Research*

Eric Cross, *Evan Pugh Professor of Electrical Engineering, The Pennsylvania State University*
P.S. Nicholson, *Professor, Department of Materials Science & Engineering, McMaster University*
Harry C. Gatos, *Professor of Electrical Engineering Emeritus, MIT*
Marc Newkirk, *Founder Lanxide Corporation*
R. De Wames, *Manager, Rockwell International*

Closing Discussion

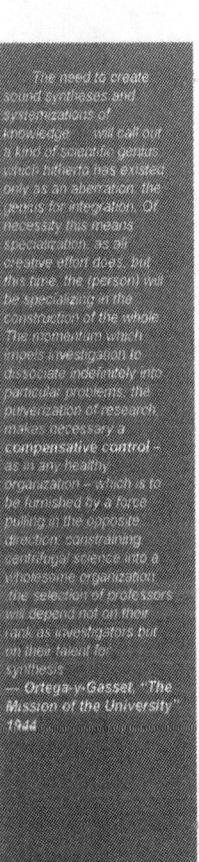

Conference Proceedings

The Conference Proceedings will consist of three parts:

Section 1:
The extended Abstracts (~2 pages each) of the remarks of each presenter

Section 2:
Recommendations by any participant to funding Agencies, and Foundations, as well as to Universities, Government Labs, and Industrial Research Units on the factors or procedures they believe can catalyze or facilitate any aspect of Interactive Research

Section 3:
Each presenter has been asked also to add any bibliography of her/his own, or others work which they regard as key texts. This will include all longer papers submitted by them.

The Proceedings will be published as a book (toexcel.com) and will be made available on the web.

The Proceedings will also be sent to all registrants and distributed to major Agencies and Foundations.

IDR Organizing Committee in The Materials Research Lab

Rustum Roy, Chair
Evan Pugh Professor Emeritus of the Solid State
L. Eric Cross
Evan Pugh Professor Emeritus of Electrical Engineering
Robert E. Newnham
ALCOA Professor Emeritus of Solid State Science
Della M. Roy
Professor Emerita of Materials Science

Information:

Conference Secretary
Kathy Moir
Materials Research Laboratory
103 MRL
University Park PA 16802

Phone: (814)863-9983
Fax: (814)863-7040
Email: kathymoir@psu.edu
Or website:
www.mrl.psu.edu/calendar/IDR/IDR.htm

The need to create sound syntheses and systematizations of knowledge ... will call out a kind of scientific genius which hitherto has existed only as an aberration: the genius for integration. Of necessity this means specialization, as all creative effort does; but this time, the (person) will be specializing in the construction of the whole. The momentum which impels investigation to dissociate indefinitely into particular problems, the pulverization of research, makes necessary a compensative control — as in any healthy organization — which is to be furnished by a force pulling in the opposite direction, constraining centrifugal science into a wholesome organization ... the selection of professors will depend not on their rank as investigators but on their talent for synthesis.
— Ortega-y-Gasset, "The Mission of the University" 1944